Vaccines for Latent Viral Infections

Edited By

Liljana Stevceva

University of Texas Rio Grande Valley
School of Medicine
2102 Treasure Hills Blvd.,
Harlingen TX
USA

DEDICATION

I would like to dedicate this book to my parents Galaba and Risto Vitanovi who taught me that honesty, hard work and perseverance are the only way to live our lives and to my children Ilija and Risto Stevcev that always inspire me to persist and move forward.

Mom when you were studying late at night after long hours of work and taking care of us, you thought that we were sleeping but I NOTICED. Dad, all the times when you were strict but just to a student, I NOTICED; every time when you put additional effort to understand and help a student, I NOTICED, and every time you turned away any attempt of bribery even at times difficult for our family, I NOTICED. Thank you for teaching me about important things in life. I could not have asked for better parents.

The only way that we will not achieve what we strive for is if stop trying.

CONTENTS

Book Cover: Deposits of gp120 (red, Alexa fluor 555) in the lymph node during acute infection with SHIV. T cells are labelled with green fluorescence (Alexa 488 CD3). Original magnification x10. Originally published in the Journal of Immunology. Liljana Stevceva, Victor Yoon, Angela Carville, Beatriz Pacheco, Birgit Korioth-Shmitz, Keith Mansfield, and Mark C. Poznansky The Efficacy Of T Cell Immune Responses Is Reduced By SHIV-KB9 gp120 In Vivo. J. Immunol. 2008, 181 (8), 5510-21

FOREWORD

Yin – Yang (Taoistic symbol for Representation of Perfect Balance)

Definition: Latency (Present but not visible or apparent).

This book updates readers in the relatively misunderstood field of viral latency. Their major focus is the status of vaccines in the prevention and treatment of a broad array of viruses capable of persisting *via* latency. It is important to differentiate between a virus in a state of latency (*i.e.* dormancy [often the lysogenic phase of a virus life cycle]) from viruses which chronically infect cells. Dormancy is the recognition that following the initial acute infection, proliferation of the virus is terminated for a variable period (often the life span of the host). When it does "awake", the lytic part of the cycle is activated causing replication and shedding, and thus reinfection.

The Chinese symbol above can be applied to the two types of viral latency – proviral *versus* episomal viral genetic material. Viral RNA or DNA hibernates in either the cytoplasm or nucleus as a foreign "immigrant". This state is vulnerable to the host's gene degradation and/or ribosomes. The fascinating aspect of how episomal viruses avoid the intracellular immune attack is that they remain outside the nucleus, which greatly reduces triggering an interferon attack *via* the Nuclear Domain 10. But the flip side of this "yin" is that these viruses are still subjected to degradation by cytoplasmic enzymes.

The other side of latency (the yang) is to the proviral universe of viral DNA integrated into the nuclear "stuff" of its host. These viruses enjoy a protected and cozy life intertwined within the host DNA and can only be eliminated by killing the entire cell. They require packaging proteins for entering the nuclear haven, which compromises latency over time.

The authors address the respective mechanisms of maintaining latency for both types of viral states, and the influence of the envelope glycoproteins for sustaining latency.

These latent viruses have different relationships in the host which range from symbiosis to outright parasitism, leading to a variety of clinical consequences. The authors describe the effectiveness of vaccines in terminating or ameliorating these infections that affect most life forms. Only partial explanations are available to what do these integrated latent virus contribute to the host as large areas of RNA/DNA have integrated viral sequences. The yin is they are present and tolerated – the yang has led to many speculations that they influence immunity, anthropomorphic textures, organogenesis, personality, genetics of the host, behavior *etc*. Are they overall beneficial (the yin) or deleterious (the yang). If the latter, why is the persistence so long? More effective vaccines against these latent viruses may eventually provide insights to the ultimate consequences of viral latency by demonstrating total prevention against latency or even eliminating viral latency form the host.

Joseph Silva
Founding Dean
Vice President for Medical Affairs
Professor of Internal Medicine and Infectious Diseases
California Northstate University College of Medicine
Dean Emeritus, UC Davis School of Medicine
USA

PREFACE

I now believe that the first buds about writing this book germinated fifteen years ago when I started seriously looking into the structure and function of the HIV gp120 glycoprotein. I became aware of the unique immunomodulating properties of sulfated polysaccharides during my PhD studies at the Australian National University. As soon as my research shifted to HIV and HIV vaccine development, the similarity of the way the envelope of the HIV was constructed to these compounds attracted my attention and I began wondering how much of the immunomodulating properties do the envelope possess and how the virus uses it. Significant scientific effort was devoted in the past 20 years to dissecting the structure of the HIV envelope glycoproteins but it still appears that our thinking about the HIV envelope did not take into account sufficiently its function and usefulness for the virus. Bearing in mind that all of the known HIV vaccine candidates had envelope component in them, this might have significantly affected their efficacy. As we are now embarking into a new era in HIV vaccine development, perhaps it is pertinent to revisit the basic immunomodulatory role of the HIV glycoproteins and mechanisms of viral escape as they relate to HIV vaccines.

My research interests recently widened to include other viruses that are capable of evading the immune response and establishing a state of latency. As I began embarking deeper into researching these viruses, it soon became very obvious to me that the envelopes of all of them contain similar structures that allow them to modulate the immune response. It also became apparent that any past or present vaccine development effort related to these viruses is directed towards developing vaccine that will prevent clinical manifestation *i.e.* disease (or what we are able to detect of it) but that there was never an effort to prevent establishment of dormant viral state. Perhaps the reason for this was that until recently, the presence of latent viruses was largely underestimated and that it was widely believed not to cause any harm to the human body as long as the immune system is intact. Novel findings from the human genome sequencing as well as emerging findings on the potential role of CMV infection in immunosenescence are however revealing that this might not be the case. It is time to have a second look at the viruses that are capable of causing latent infection and to shift the scientific taught towards preventing establishment of latency.

Liljana Stevceva,
University of Texas Rio Grande Valley
School of Medicine
2102 Treasure Hills Blvd.,
Harlingen TX
USA

CONTRIBUTORS

Barghavi Patham, MD, PhD   Baylor College of Medicine Endocrinology, 1504 Taub Loop Houston, TX 77030, USA

Ghaith Al Eyd, MD, PhD   California Northstate University College of Medicine, 9700 West Taron Drive, Elk Grove, CA, USA

Kallie Rebekah Appleton, MD   The Paul Foster School of Medicine, TexasTech, 3068 Hayden Road, Columbus, Ohio 43235, USA

Liljana Stevceva, MD, PhD   University of Texas Rio Grande Valley School of Medicine, 2102 Treasure Hills Blvd., Harlingen TX, USA

Masha Fridkis-Hareli, PhD   ATR LLC, Sudbury, MA 01776, USA

Nelly Estrada, MD   The Paul Foster School of Medicine, TexasTech, 5001 El Paso Drive, El Paso, TX 79905, USA

Risaku Fukumoto, PhD   Scientific Research Department, Armed Forces Radiobiology Research Institute, Uniformed Services University of the Health Sciences, USA

Sandesh Subramanya, PhD   Bioo Scientific Corp., 7050 Burleson Rd. Austin, TX 78744, USA

Tu Thanh Mai, MD   The Paul Foster School of Medicine, TexasTech, 5001 El Paso Drive, El Paso, TX 79905, USA; 2401 S 31ST ST, Temple, TX 76508, USA

CHAPTER 1

Latent Viral Infections in Humans

Liljana Stevceva[*]

California Northstate University College of Medicine, 9700 West Taron Drive, Elk Grove, CA 95757, USA

Abstract: Some viruses are able to escape the immune responses generated against them and to establish a latent state that is not visible to the immune system. Such viruses hide in CD4+ T cells, all T cells, B lymphocytes, germinal epithelial cells, neurons and others. They can reactivate at times when the immunity is compromised. It is not known what are the effects, if any, of dormant viruses on the immune system. Some of these viruses contribute to malignant transformation of cells. Recent studies have implicated that presence of latent viruses such as CMV might be driving the T cells to terminal differentiation and exhaustion. Retroviruses incorporate into the human genome and now represent as much as 8% of it constituting of about 30 000 different retroviruses. In addition, various other transposable elements like these make up to 45% of the human genome. Other latent viruses such as Bornavirus are believed to be involved in the pathogenesis of human psychiatric diseases such as bipolar disorder and depression. Attempts to design vaccines that will prevent the virus from going into latency have been scarce.

Keywords: Immune evasion, latent infection, persistent infection, dormant, reactivation, CD4+T cell, HIV, HHV-6, HHV-7, CMV, HTLV-1, HTLV-2, HSV, VZV, EBV, JC, VK, measles, immunosenesence, terminally differentiated, endogenous retroviruses, retrotransposons, prokaryotic retrons, "retrotranscripts", CMV, Bornavirus,

In most of the cases infection with viruses induces immune responses that eventually destroy the virus and virus-infected cells and result in clearing the infection. Sometimes after the initial viral infection, the virus establishes equilibrium with the host and it is not destroyed by the immune system but it remains in a dormant, latent state where it is invisible to the body defense systems. Such, latent virus does not replicate and cannot be detected but it can be reactivated and can begin to replicate again. Reactivation usually occurs in response to external stimuli and without the need of reinfection with new viruses. The virus usually prefers certain cells or tissues where it goes into latent state

*Corresponding author Liljana Stevceva: University of Texas Rio Grande Valley School of Medicine, 2102 Treasure Hills Blvd., Harlingen TX, USA; E-mail: Liljana@hotmail.com

therefore creating reservoirs of latent infection. For many of the viruses capable of going into latency the preferred cell is the CD4+ T cell (HIV, HHV-6, HHV-7, CMV) or all T cells (HTLV-1 and HTLV-2, CMV). EBV hides in B lymphocytes, HPV in germinal epithelial cells while HSV-1, HSV-2, VZV, JC, VK and measles are harbored in neurons and other cells of the CNS [reviewed in (Boldogh *et al.*, 1996)].

Within such reservoirs the virus can remain in latent state indefinitively so, latent viral infections are a type of persistent infection. Viruses may also establish persistent chronic infections in which case the virus is replicating and can be detected in the body. These two states can interchange and the virus occasionally reactivates from the latent state, starts replicating (during which time the carrier is asymptomatic and the virus could spread to other people) and goes back to latent state without ever being completely cleared from the body. Thus, persistent viral infections are those in which the virus is not cleared after the primary infection but remains associated with cells in the body (Boldogh *et al.*, 1996). Persistent viral infections can be established as latent infections where the virus is dormant, is not replicating and cannot be detected in the body or as chronic infections where virus can be detected. The later can manifest as chronic or recurrent disease or as slow infection where there is a prolonged incubation period followed by progressive disease.

Traditionally, very little attention has been paid to the latent phase of the viral infection and that goes for the mechanisms that allow the virus to remain dormant and undetected by the body's immune system and for the mechanisms that induce reactivation. Many of the Herpesviruses have the capacity to establish latent infections (HSV, HPV, CMV, VZV *etc*) but so do retroviruses such as HIV and HTLV. And while we do know that some viruses have the ability to escape the immune recognition and the immune response and can establish latent infection we know very little about the structural properties of these viruses that perhaps skew the immune response. Do they have anything in common? Do they have structures, receptors, secrete proteins that are similar or have similar mechanisms to evade the immune responses? Even less is known about the possible effects of such latent viruses on the body immune system. Do they influence the way the immune system functions and if yes, how? Or is the virus just hiding in the body undetected and without having any effect on the immune system? Till recently, the latest was widely believed. This is why latent infections did not receive their due attention from the scientific community. But are they really harmless? It is well known that latent infection with HTLV, EBV and HPV for example contributes to malignant transformation of cells and can lead to tumors but the exact mechanisms of this transformation, when and in which cases does it occur are yet to be fully understood.

As a matter of fact, several studies on CMV and immunosenescence have indicated that CMV infection significantly influences T cell responses especially in the elderly, a population that is about 80% seropositive for the virus. While it was known for a while that CMV infection causes changes in T cell populations (Looney *et al.*, 1999), the advancement in technology and availability of multiparametar flowcytometry recently made it possible to better analyze these T cell subsets. Infection with CMV in asymptomatic individuals was associated with significantly higher number of terminally differentiated CD4+ and CD8+ T cells and lower frequency of naïve CD8+ T cells (Derhovanessian *et al.*, 2011). It seems logical that the accumulation of such terminally differentiated T cells that are short lived at the expense of the naïve T cell pool would affect T cell responses to viral infections, vaccines and tumors. Thus, it comes as no surprise that a recent study reported that accumulation of terminally differentiated CD4+ T cells in CMV infected asymptomatic individuals over 60 years of age correlated with poor response to influenza vaccine (Derhovanessian *et al.*, 2013). It also seems highly unlikely that anything that affects the CD4+ T cell response will not affect B cell help and humoral responses to protein antigens and it will be interesting to see what other findings come out. As the CMV story unravels, the scientific community begins to question the influence of other latent viral infections on the immune system. As this happens we need to keep an open mind to the various immune compartments that might be affected in a variety of ways.

Viral latency may be established by episomal latency mechanism where the viral DNA persists within the nucleus or the cytoplasm of the host cell (common in Herpesviruses) or by proviral latency mechanism where the viral gene inserts itself into the host genome and creates a provirus (usually Retroviruses). Some scientists still argue that HIV does not establish 'true latency' because it can be detected at low levels with modern methods in the infected asymptomatic individuals, thus, it is replicating. However, early studies done *in vitro* have shown that the virus is latent at least in pure CD4+ T cells cultures. Here, HIV can be reactivated if the CD4+ T cells become active and the virus is then rescued and can be detected by PCR (Chun *et al.*, 1995). This could explain that the low levels of virus still detected in the blood of asymptomatic patients are simply a consequence of the regular activation processes of CD4+T cells. In addition, it has been known for a while that various retroelements such as endogenous retroviruses, the retrotransposons, the prokaryotic retrons and the "retrotranscripts" (*Alu*-like sequences and processed pseudogenes) can be found incorporated in the human and animal genome (Boeke and Stoye, 1997). The sheer magnitude of this phenomenon was unknown until the sequencing of the full human genome was completed and it became clear that 45% of the human genome consists of transposable elements like these (Lander *et al.*, 2001), (Venter *et al.*, 2001). If the possibility that many such elements could have degenerated to a non detectable level is calculated in, it means that over 50% of the human genome might have originated from insertion of repetitive elements (Li *et al.*, 2001). About 8% of human genome consists of sequences that are easily recognizable as coming from infectious retroviruses (Griffiths, 2001). Research determining what is the effect of these retroelements to human health and disease is yet to be done. In 2009, japanese researchers published a study identifying elements homologous to the nucleoprotein of Bornavirus, a non-retrovirus that belongs to the family *Mononegavirales* and infects birds and mammals (Horie *et al.*, 2010). Bornavirus got its name by the german town Borna where it caused epidemic among a regiment of cavalry horses in 1885. Bornavirus is a neurotropic virus that establishes persistent infection in the host brain and it has been implicated in the pathogenesis of human psychiatric disorders such as bipolar disorder and depression (Rott *et al.*, 1991), (Miranda *et al.*, 2006), (Fukuda *et al.*, 2001). It is believed that Bornavirus particles became incorporated in the human brain about 40 million years ago.

As previously mentioned, once the virus goes into latency, it becomes invisible to the immune defense mechanisms and it cannot be detected. Problem is that such latent viruses can persist for a very long time. They can reactivate into an acute infection, go on and off into persistent chronic infection or even worse, they can transform and push the cell in which they reside into uncontrolled cell division. This cell division will lead to the viral gene being inserted into the host genome. Such was the case in a study using retroviral vectors to deliver gene therapy for a genetic disorder where five of the 20 participants developed leukemia-like symptoms (Hacein-Bey-Abina *et al.*, 2003), (Hacein-Bey-Abina *et al.*).

Because of the widely held belief that they are harmless, research efforts focused on eradicating latent and persistent chronic viral infection have been extremely scarce and no approaches have been developed to date to target this problem. As a matter of fact, vaccines and therapeutic approaches against viruses that go into latency are focused on minimizing the clinical manifestations of the viral infections. Immunization strategies employed so far have not been very successful. A novel approach is needed that takes into account the immune escape mechanisms that allow the virus to go into latency. If these viruses are prevented from escaping the immune system and are forced to remain visible to it, they will be targeted and destroyed at the beginning of the infection.

CONFLICT OF INTEREST

The author confirms that this chapter contents have no conflict of interest.

ACKNOWLEDGEMENTS

Declared None.

REFERENCES

BOEKE, J. D. & STOYE, J. P. 1997. Retrotransposons, Endogenous Retroviruses, and the Evolution of Retroelements. *In:* COFFIN, J. M., HUGHES, S. H. & VARMUS, H. E. (eds.) *Retroviruses.* Cold Spring Harbor (NY).

BOLDOGH, I., ALBRECHT, T. & PORTER, D. D. 1996. Persistent Viral Infections. *In:* BARON, S. (ed.) *Medical Microbiology.* 4th ed. Galveston (TX).

CHUN, T. W., FINZI, D., MARGOLICK, J., CHADWICK, K., SCHWARTZ, D. & SILICIANO, R. F. 1995. *In vivo* fate of HIV-1-infected T cells: quantitative analysis of the transition to stable latency. *Nat Med,* 1, 1284-90.

DERHOVANESSIAN, E., MAIER, A. B., HAHNEL, K., BECK, R., DE CRAEN, A. J., SLAGBOOM, E. P., WESTENDORP, R. G. & PAWELEC, G. 2011. Infection with cytomegalovirus but not herpes simplex virus induces the accumulation of late-differentiated CD4+ and CD8+ T-cells in humans. *J Gen Virol,* 92, 2746-56.

DERHOVANESSIAN, E., THEETEN, H., HAHNEL, K., VAN DAMME, P., COOLS, N. & PAWELEC, G. 2013. Cytomegalovirus-associated accumulation of late-differentiated CD4 T-cells correlates with poor humoral response to influenza vaccination. *Vaccine,* 31, 685-90.

FUKUDA, K., TAKAHASHI, K., IWATA, Y., MORI, N., GONDA, K., OGAWA, T., OSONOE, K., SATO, M., OGATA, S., HORIMOTO, T., SAWADA, T., TASHIRO, M., YAMAGUCHI, K., NIWA, S. & SHIGETA, S. 2001. Immunological and PCR analyses for Borna disease virus in psychiatric patients and blood donors in Japan. *J Clin Microbiol,* 39, 419-29.

GRIFFITHS, D. J. 2001. Endogenous retroviruses in the human genome sequence. *Genome Biol,* 2, REVIEWS1017.

HACEIN-BEY-ABINA, S., VON KALLE, C., SCHMIDT, M., LE DEIST, F., WULFFRAAT, N., MCINTYRE, E., RADFORD, I., VILLEVAL, J. L., FRASER, C. C., CAVAZZANA-CALVO, M. & FISCHER, A. 2003. A serious adverse event after successful gene therapy for X-linked severe combined immunodeficiency. *N Engl J Med,* 348, 255-6.

HORIE, M., HONDA, T., SUZUKI, Y., KOBAYASHI, Y., DAITO, T., OSHIDA, T., IKUTA, K., JERN, P., GOJOBORI, T., COFFIN, J. M. & TOMONAGA, K. 2010. Endogenous non-retroviral RNA virus elements in mammalian genomes. *Nature,* 463, 84-7.

LANDER, E. S., LINTON, L. M., BIRREN, B., NUSBAUM, C., ZODY, M. C., BALDWIN, J., DEVON, K., DEWAR, K., DOYLE, M., FITZHUGH, W., FUNKE, R., GAGE, D., HARRIS, K., HEAFORD, A., HOWLAND, J., KANN, L., LEHOCZKY, J., LEVINE, R., MCEWAN, P., MCKERNAN, K., MELDRIM, J., MESIROV, J. P., MIRANDA, C., MORRIS, W., NAYLOR, J., RAYMOND, C., ROSETTI, M., SANTOS, R., SHERIDAN, A., SOUGNEZ, C., STANGE-THOMANN, N., STOJANOVIC, N., SUBRAMANIAN, A., WYMAN, D., ROGERS, J., SULSTON, J., AINSCOUGH, R., BECK, S., BENTLEY, D., BURTON, J., CLEE, C., CARTER,

N., COULSON, A., DEADMAN, R., DELOUKAS, P., DUNHAM, A., DUNHAM, I., DURBIN, R., FRENCH, L., GRAFHAM, D., GREGORY, S., HUBBARD, T., HUMPHRAY, S., HUNT, A., JONES, M., LLOYD, C., MCMURRAY, A., MATTHEWS, L., MERCER, S., MILNE, S., MULLIKIN, J. C., MUNGALL, A., PLUMB, R., ROSS, M., SHOWNKEEN, R., SIMS, S., WATERSTON, R. H., WILSON, R. K., HILLIER, L. W., MCPHERSON, J. D., MARRA, M. A., MARDIS, E. R., FULTON, L. A., CHINWALLA, A. T., PEPIN, K. H., GISH, W. R., CHISSOE, S. L., WENDL, M. C., DELEHAUNTY, K. D., MINER, T. L., DELEHAUNTY, A., KRAMER, J. B., COOK, L. L., FULTON, R. S., JOHNSON, D. L., MINX, P. J., CLIFTON, S. W., HAWKINS, T., BRANSCOMB, E., PREDKI, P., RICHARDSON, P., WENNING, S., SLEZAK, T., DOGGETT, N., CHENG, J. F., OLSEN, A., LUCAS, S., ELKIN, C., UBERBACHER, E., FRAZIER, M., *et al.* 2001. Initial sequencing and analysis of the human genome. *Nature,* 409, 860-921.

LI, W. H., GU, Z., WANG, H. & NEKRUTENKO, A. 2001. Evolutionary analyses of the human genome. *Nature,* 409, 847-9.

LOONEY, R. J., FALSEY, A., CAMPBELL, D., TORRES, A., KOLASSA, J., BROWER, C., MCCANN, R., MENEGUS, M., MCCORMICK, K., FRAMPTON, M., HALL, W. & ABRAHAM, G. N. 1999. Role of cytomegalovirus in the T cell changes seen in elderly individuals. *Clin Immunol,* 90, 213-9.

MIRANDA, H. C., NUNES, S. O., CALVO, E. S., SUZART, S., ITANO, E. N. & WATANABE, M. A. 2006. Detection of Borna disease virus p24 RNA in peripheral blood cells from Brazilian mood and psychotic disorder patients. *J Affect Disord,* 90, 43-7.

ROTT, R., HERZOG, S., BECHTER, K. & FRESE, K. 1991. Borna disease, a possible hazard for man? *Arch Virol,* 118, 143-9.

VENTER, J. C., ADAMS, M. D., MYERS, E. W., LI, P. W., MURAL, R. J., SUTTON, G. G., SMITH, H. O., YANDELL, M., EVANS, C. A., HOLT, R. A., GOCAYNE, J. D., AMANATIDES, P., BALLEW, R. M., HUSON, D. H., WORTMAN, J. R., ZHANG, Q., KODIRA, C. D., ZHENG, X. H., CHEN, L., SKUPSKI, M., SUBRAMANIAN, G., THOMAS, P. D., ZHANG, J., GABOR MIKLOS, G. L., NELSON, C., BRODER, S., CLARK, A. G., NADEAU, J., MCKUSICK, V. A., ZINDER, N., LEVINE, A. J., ROBERTS, R. J., SIMON, M., SLAYMAN, C., HUNKAPILLER, M., BOLANOS, R., DELCHER, A., DEW, I., FASULO, D., FLANIGAN, M., FLOREA, L., HALPERN, A., HANNENHALLI, S., KRAVITZ, S., LEVY, S., MOBARRY, C., REINERT, K., REMINGTON, K., ABU-THREIDEH, J., BEASLEY, E., BIDDICK, K., BONAZZI, V., BRANDON, R., CARGILL, M., CHANDRAMOULISWARAN, I., CHARLAB, R., CHATURVEDI, K., DENG, Z., DI FRANCESCO, V., DUNN, P., EILBECK, K., EVANGELISTA, C., GABRIELIAN, A. E., GAN, W., GE, W., GONG, F., GU, Z., GUAN, P., HEIMAN, T. J., HIGGINS, M. E., JI, R. R., KE, Z., *KET*CHUM, K. A., LAI, Z., LEI, Y., LI, Z., LI, J., LIANG, Y., LIN, X., LU, F., MERKULOV, G. V., MILSHINA, N., MOORE, H. M., NAIK, A. K., NARAYAN, V. A., NEELAM, B., NUSSKERN, D., RUSCH, D. B., SALZBERG, S., SHAO, W., SHUE, B., SUN, J., WANG, Z., WANG, A., WANG, X., WANG, J., WEI, M., WIDES, R., XIAO, C., YAN, C., *et al.* 2001. The sequence of the human genome. *Science,* 291, 1304-51.

CHAPTER 2

Characteristics of Viruses that Induce Latent Infections

Bhargavi Patham[1,2], Nelly Estrada[2] and Sandesh Subramanya[1,*]

[1]*Departments of Biomedical Sciences and* [2]*Internal Medicine, Paul L. Foster School of Medicine, Texas Tech University Health Sciences Center, 5001 El Paso Dr, El Paso, TX 79912, USA*

Abstract: In latent infections, overt disease is not produced, but the virus is not eradicated. The virus may exist in a truly latent noninfectious occult form, possibly as an integrated genome or an episomal agent, or as an infectious and continuously replicating agent, termed as a persistent chronic viral infection. Reactivation of a latent infection may be triggered by various stimuli, including changes in cell physiology, super-infection by another virus, and physical stress or trauma. Viruses that cause latent infections can persist at the same time in different cell types of one or more tissues or organs. This review summarizes the characteristics and mechanisms by which persistent infections are maintained by modulation of both virus and cellular gene expression and modification of the host immune response.

Keywords: DNA viruses, Latency, reactivation, herpesvirus, papillomavirus, retrovirus, HIV, EBV, VZV, HCV, HPV, HCMV, latency associated transcripts (LATS), major IE promoter/enhancer (MIEP), EBV nuclear antigen (EBNA1), Burkitt's lymphoma, concatamers, dorsal root ganglion (DRG), CD4, CCR5, gut-associated lymphoid tissue (GALT), HAART.

INTRODUCTION

Most viruses cause acute, self-limiting disease whereby the pathogen replicates rapidly intracellularly and disseminates to another organism prior to immune clearance or death of the host. However, some viruses are able to establish a persistent infection in the host by manipulating cellular mechanisms for their own advantage. Viral latency can be defined as the maintenance of viral genome for the lifetime of the host after primary infection with the ability to reactivate under certain conditions. Persistence can occur through the absence of productive

*Corresponding author Sandesh Subramanya: Department of Biomedical Science, Paul L. Foster School of Medicine, Texas Tech University Health Sciences Center, 5001 El Paso Dr, El Paso, TX 79912, USA; E-mail: sunsand@gmail.com

Liljana Stevceva (Ed)

infectious virions, proviral integration into the host genome, and/or continuous viral replication. Although each virus has evolved to enable it to undergo latency there are similarities in the establishment and maintenance of long-term infections. These include (i) viral selection of cellular subsets for latent infection, (ii) modulation in expression of viral genes, (iii) avoidance of host immune surveillance, (iv) viral subversion of detrimental apoptotic pathways, and (v) reactivation upon opportunistic infections or in an immunosuppressed or immunocompromised host.

MAJOR VIRAL FAMILIES IN THE MUCOSAL CAVITY THAT INDUCE LATENT INFECTIONS

Primary viral infection begins when the pathogen interacts with mucosal surfaces of the respiratory, genital, or gastrointestinal tracts. Some viral infections are limited to the mucosa; in many others virus crosses the mucosal surface and spreads to other organs (Kane and Golovkina, 2010). The virus has to overcome several mucosal barriers like viscous fluids, membrane-bound glycocalyx, intercellular tight junctions, mucosal immunoglobulin, anti-viral cellular components before being able to invade the sub-epithelial layer and traverse to secondary sites.

Of the DNA viruses, members belonging to the family Herpesviridae cause persistent infection (Table **1**). Members of the Herpesviridae (EBV, VZV, HSV-1 and CMV, HSV-2) have characteristic morphology with an icosahedral capsid. The capsid is surrounded by protein structured tegument which in turn is encapsulated by an envelope that consists of glycoproteins, polyamines, and glycolipids The glycoproteins confer distinctive properties to each virus and provide the antigens to which the host is capable of responding. Each herpes virus subfamily maintains latent infection in specific cell population (s). Of the RNA viruses HIV-1, HTLV-1 and HTLV-2 cause latent infection in their host. Below we describe important viruses that are initially associated with the mucosal surface but eventually lead to persistent infections.

Herpes Simplex Virus (HSV-1 and HSV-2)

HSV-1 and HSV-2 belong to the sub-family *Alphaherpesvirinae* and are genetically related with about 70% homology. In addition to DNA composition these two viruses can be distinguished by antigenic expression. Both viruses infect the mucosal surfaces with HSV-1 predominantly responsible for the

Table 1. Characteristics of viruses that cause chronic infections.

	Formal Abbreviation	Genome (size)	Capsid	Envelope
DNA Virus				
Alphaherpesviridae				
Simplexvirus				
Herpes Simplex Virus-1 (**HSV-1**)	HHV-1	ds (152 kb)	Icosahedral	Yes
Herpes Simplex Virus-2 (**HSV-2**)	HHV-2	ds (155 kb)	Icosahedral	Yes
Variocellosis				
Varicella-zoster Virus (**VZV**)	HHV-3	ds (125 kb)	Icosahedral	Yes
Betaherpesvirinae				
Cytomegalovirus (**HCMV**)	HHV-5	ds (235 kb)	Icosahedral	Yes
Gammaherpesvirinae				
Lymphocryptovirus				
Ebstein Barr Virus (**EBV**)	HHV-4	ds (172 kb)	Icosahedral	Yes
Papillomaviridae				
Human Papilloma Virus (**HPV**)	HPV	ds (8 kb)	Icosahedral	No
RNA Virus				
Lentivirus				
Human Immunodeficiency virus (**HIV-1**)	HIV-1	ss (7-11 kb)	Icosahedral	Yes

ds: double-stranded; ss: single-stranded; kb: kilobases

majority of oral lesions and fewer cases where sexual contact is the route of transmission (Kutok and Wang, 2006, Whitley, 2002). HSV-2 infection is traditionally associated with sexual route of transmission and more commonly produces genital ulcerations rather than oral lesions (Looker *et al.*, 2008). This infection classically occurs in young adulthood through intimate mucocutaneous contact and exposure to the virus (Jones, 2003, Nahmias and Roizman, 1973), (Roizman and Whitley, 2013). In addition, severe morbidity may result from recurrent genital herpes in immunocompromised patients. HSV-2 also appears to enhance the risk of acquisition of HIV by 2- to 3-fold. However the division between HSV-1 and 2 infections in the respective sites is not clear-cut, and HSV-2 infection is, on occasion, responsible for oral ulcers.

Following invasion *via* the oral or genital mucosa HSV-1 primarily replicates in stratified squamous epithelial cells (Cunningham *et al.*, 2006) (Table **2**). Active infection may be asymptomatic but it still gives rise to viral shedding or can cause fever, sore throat, and painful vesicles, often at a posterior location. In adolescents, and to a lesser extent in adults, it may frequently cause pharyngitis together with lesions on the buccal and gingival mucosa (herpetic gingivo-stomatitis). In severe cases, dysphagia and lymphadenopathy may be present. In an immunocompetent patient the episode usually resolves spontaneously within 10–14 days. During primary infection, the virus migrates into ramifying unmyelinated sensory nerve fibers within the stratified squamous epithelium and eventually retrogrades to the cell body of the neuron in the dorsal root ganglion (DRG) adjacent to the spinal cord (or the trigeminal ganglion for HSV-1) where it establishes latent infection in a nonintegrated (episomal) circular/ concatemeric form (Perng *et al.*, 2000, Stevens *et al.*, 1987, Thompson and Sawtell, 2000), (Lanfranca *et al.*, 2014). During latency, restricted HSV-1/2 gene expression produces two collinear latency associated transcripts (LATs) that are thought to bind to polyribosomes and maintain transcriptional and translational silencing of the viral genome (Kent *et al.*, 2003, Perng *et al.*, 2000, Efstathiou and Preston, 2005), (Watson *et al.*, 2013). It has been suggested that anti-apoptotic properties of LAT are important in the latency-reactivation cycle (Henderson *et al.*, 2002, Perng *et al.*).

Varicella Zoster Virus (VZV)

VZV is an exclusively human, highly neurotropic alphaherpesvirus. Primary infection causes chickenpox (varicella) that is typically seen in children 1 to 9 years of age (Finger *et al.*, 1994, Kudesia *et al.*, 2002), (Baird *et al.*, 2013). Primary infection produces 4 million cases of chickenpox annually. In temperate climates more than 90% of people are infected before adolescence with an incidence of 13–16 cases per 1000, with peak incidence during winter and spring (Kudesia *et al.*, 2002). VZV becomes latent in neurons of the cranial nerve ganglia, dorsal root ganglia, and autonomic ganglia along the entire neuraxis (Gilden *et al.*, 1987, Mahalingam *et al.*, 1990). Decades later, VZV may reactivate in the elderly and the immuno-compromised to cause herpes zoster (shingles), pain, and rash. Zoster affects approximately 1 million individuals in the United States per year. The annual incidence of zoster is approximately 5-6.5/1000 individuals at age 60, increasing to 8-11/1000 at age 70 (Kilgore *et al.*, 2003, Donahue *et al.*, 1995).

Table 2. Primary and latent cells infected by chronic or latent viruses.

Virus	Abbreviation	Primary	Latent
Human Immunodeficiency virus	HIV	CD4+ T cell	CD4+ T cell
Human Herpes Virus (1,2)	HSV (1,2)	Muco-epithelial cells	Neurons
Human Herpes Virus 3	VZV	CD4+ T cell, Muco-epithelial cells	Neurons
Human Herpes Virus 4	EBV	Epithelial cells, B cells	B cells
Human Herpes Virus 5	CMV	Epithelial cells, monocytes, lymphocytes	Monocytes, Lymphocytes

Latent VZV DNA is circular or concatameric (end-to-end) state and present as 30 to 3500 VZV DNA copies per 100 ng of total ganglionic DNA (Mahalingam *et al.*, 1992). VZV genome consists of a unique long (U_L; 104 kbp) and a unique short (U_S; 0.5 kbp) DNA segments bound by inverted repeats (Davison and Scott, 1986). Transcripts and proteins corresponding to VZV genes 21, 29, 62, 63, and 66 have been identified in latently infected human (LaGuardia *et al.*, 1999, Cohrs and Gilden, 2003), (Zerboni and Arvin, 2011). These proteins are localized in the cytoplasm of latently infected neurons of dorsal root ganglia and upon reactivation localize the nucleus (Kennedy and Cohrs, 2010), (Nagel and Gilden, 2014). Some of these VZV proteins may restrict regulatory proteins from entering the nucleus, thereby maintaining a latent state in the presence of the essential immediate-early protein product of ORF62, an efficient transactivator of VZV gene transcription (Moriuchi *et al.*, 1994, Moriuchi *et al.*, 1995, Eisfeld *et al.*, 2006). The protein kinase 66-pk, encoded by ORF66, phosphorylates the protein product of ORF62, leading to its cytoplasmic accumulation (Eisfeld *et al.*, 2006).

Cytomegalovirus (CMV)

CMV is a ubiquitous herpes virus and the largest as compared to other human herpes viruses, with a genome of 235 kb encoding 165 genes (Crough and Khanna, 2009). The virion consists of a double-stranded linear DNA genome core in an icosahedral nucleocapsid, enveloped by a proteinaceous matrix (the tegument). The tegument proteins include the lower matrix phosphoprotein 65 (pp65), upper matrix transactivator phosphoprotein 71 (pp71), and the large matrix virion maturation phosphoprotein 150 (pp150) (Yurochko, 2008, Kalejta, 2008b). Additionally, the tegument also contains cellular and viral RNA (Landolfo *et al.*, 2003). The tegument proteins play an important role in virion structure and assembly and modulate host cell response to infection (Kalejta, 2008a), (Schleiss *et al.*, 2013). The viral envelope is derived from the host cell endoplasmic reticulum-Golgi-derived lipid bilayer envelope and contains at least 20 virus-encoded glycoproteins that are involved in cell attachment and penetration including glycoprotein B (gB), gH, gL, gM, gN, and gO (Castillo and Kowalik, 2002, Landolfo *et al.*, 2003). Viral replication is controlled by expression of immediate early (IE), early (E), and late (L) phase genes in overlapping phases (Liu *et al.*, 2013).

Primary infection is asymptomatic in an immunocompetent host despite the high frequency of CMV specific antibodies. Clinical disease is more likely in the immunosuppressed, particularly in transplant recipients and individuals with AIDS. In fact 75% of the organ transplant recipients are affected by CMV infection during the first year after transplantation. IE gene products are particularly important in the pro-inflammatory response seen in CMV infection and may (a) increase levels of adhesion molecules such as ICAM, (b) increase production of cytokines such as IL-1 and TNF-α; (c) upregulate several chemokines like MCP-1, MIP-1β, RANTES, and IL-8; and (d) stimulate expression of growth factors and mitogens including IL-6, TGF-β and GM-CSF (Scholz *et al.*, 2003, Colberg-Poley, 1996). CMV infection of monocytes, endothelial cells, and smooth muscle cells is central to the pro-inflammatory process. CMV subverts immune surveillance by several mechanisms including reducing the apoptosis of infected cells, blocking antigen presentation, and sequestering chemokines from extracellular environment (Vancikova and Dvorak, 2001, Bodaghi *et al*., 1998, Sedmak *et al*., 1994, Varani *et al*., 2005), (McCormick and Mocarski, 2015).

Latent HCMV DNA is detected in $CD34^+$ hematopoetic stem cells, which are progenitors for cells of B, T, and monocyte/macrophage lineages, and in monocytes, but not in mature B or T cells (Taylor-Wiedeman *et al*., 1994, Mendelson *et al*., 1996, Larsson *et al*., 1998b, Larsson *et al*., 1998a). Three important pathways are thought to lead to establishment of latency (Sinclair and Sissons, 2006, Sissons *et al*., 2002). First following initial attachment and entry, HCMV may enter directly into a latent state without *de novo* viral gene expression. Second, a virus-initiated productive infection may be interrupted prematurely, subsequently leading to latency. Third, following entry, a subset of viral genes not associated with productive infection maybe required for the successful establishment of latency. An antisense transcript to viral UL82 gene that encodes the viral pp71 tegument protein, a known transcriptional activator of the viral major IE promoter/enhancer (MIEP has been identified in monocytes of healthy, seropositive carriers)(Nicholson *et al*., 2009). It has been suggested that this transcript or its protein product may be involved in restricting IE gene expression thereby helping to maintain latency (Goodrum *et al*., 2002, Sinclair *et al*., 1992, Bego *et al*., 2005).

Human Papilloma Virus (HPV)

HPVs are small double-stranded DNA viruses with more than 130 types (de Villiers *et al*., 2004) identified and all infect either cutaneous or squamous epithelia at specific body locations (Jung *et al*., 2004). Mucosal papillomas in the respiratory and genital-tract are predominantly induced by HPV 6, HPV 11, with other types present at lower frequencies (Gissmann *et al*., 1982, Schneider, 1993, Mounts *et al*., 1982, Maran *et al*., 1995, Stubenrauch and Laimins, 1999). HPV is associated with nearly all the cervical carcinomas of the uterine cervix, with 70% of the cases associated with the high-risk HPV types 16 and 18 (Beaudenon *et al*., 1986, zur Hausen, 2009). In most people, infection with HPV is asymptomatic and is cleared within 1 or 2 years in 90% of the cases. Genital HPV infection is considered the most common sexually transmitted disease that is responsible for more than 500,000 new annual cases of human cancer worldwide, including those of the cervix, anus, vulva, penis and of the head and neck area and oropharynx (Abramson *et al*., 2004, Doorbar, 2006), (Fiorina *et al*., 2014).

Epstein Barr Virus (EBV)

Epstein-Barr virus is considered the most effective parasite as the virus has supremely adapted to its human host and co-evolved over millions of years. Like all herpes viruses, EBV has latent and productive (lytic) phases in its life cycle. More than 90% of adults worldwide have been infected with EBV and carry the virus as a life-long persistent infection and virus production into saliva. During childhood, primary infection is spread *via* salivary contact between family members (Kutok and Wang, 2006). In adolescence, the infection precipitates as infectious mononucleosis (IM) in 50-74% of cases by kissing and the virus is found in saliva (Macsween and Crawford, 2003). For this reason IM is also called 'kissing disease'. EBV infects multiple cell lines, including B- and T-lymphocytes and squamous epithelial cells of the oropharyngeal and nasopharyngeal mucosa. Inspite of EBV's known tropism for epithelial cells, exemplified by the several EBV-associated tumors of epithelial origin, EBV- infected epithelial cells have not been identified in tonsils removed from patients with IM, whereas both latent and lytic infection of B cells can be detected (Thorley-Lawson, 2001, Thorley-Lawson, 2005).

An initial round of lytic replication occurs in tonsillar epithelium followed by infection of B cells in the lymphoepithelium of Waldeyer's ring (Hoagland, 1955). Here crypt structures that dip into underlying lymphoid tissue could allow EBV direct access to B cells. EBV displays distinct patterns of viral latent gene expression during development of both lymphoid and epithelial tumors (Young and Murray, 2003, Joseph *et al.*, 2000). Specific EBV latent gene expression patterns have been reported from different latency stages. Four major forms of EBV latent gene expression are known (Young and Murray, 2003, Billaud *et al.*, 2009, Young and Rickinson, 2004). In type 0 latency no viral genes are expressed in quiescent, memory B cells. In Type I latency, seen in Burkitt's lymphoma, there is constitutive expression of an EBV nuclear antigen, EBNA1, and small, non-polyadenylated (and therefore non-coding) RNAs, EBER1 and EBER2. EBV infection in nasopharyngeal carcinoma, Hodgkin's disease, peripheral T-cell lymphomas is the type II latency pattern, in which EBV- encoded EBNA1, EBER RNAs, and latent membrane proteins LMP-1 and LMP-2 are expressed (Babcock and Thorley-Lawson, 2000, Tsurumi *et al.*, 2005). Type III latency infection means the expressions of EBNA1, EBNA2, EBNA3, LMP1, LMP2, EBER and EBNA-LP RNAs, which are detected in patients with infectious mononucleosis (IM), post-transplantation lympho-proliferative disorders and acquired immunodeficiency syndrome-related lymphomas (Thompson and Kurzrock, 2004,

Pang *et al.*, 2009, Babcock and Thorley-Lawson, 2000), (Bentz *et al.*, 2015), (Lai *et al.*, 2015).

Lytic replication differs from the latent state in that multiple rounds of replication are initiated within oriLyt (Origin of Lytic) locus. Upon induction of the lytic program two key EBV immediate-early (IE) lytic genes, BZLF1 and BRLF1 are expressed (Billaud *et al.*, 2009, Holley-Guthrie *et al.*, 1990, Kenney *et al.*, 1989b). These genes encode transactivators that activate viral and certain cellular promoters leading to an ordered cascade of viral gene expression (Kenney *et al.*, 1989a). They activate quiescent B cells to enter into the cell cycle, maintaining continuous proliferation, and prevent them from undergoing apoptosis. Additionally lytic viral proteins also include an IL-10 homologue that inhibits co-stimulatory and antigen-presentation functions of monocytes/macrophages (Stuart *et al.*, 1995, Stewart and Rooney, 1992) and a Bcl2 homologue that prolongs cell survival by inhibiting apoptosis (Oudejans *et al.*, 1995, D'Sa-Eipper *et al.*, 1996). These immune evasion strategies buy time for effective production of viral particles.

Human Immunodeficiency Virus (HIV)

HIV is a retrovirus, a member of the *Lentivirus* genus, consisting of a cylindrical nucleocapsid enclosing a RNA genome and a reverse transcriptase enzyme that produces a DNA copy of the RNA genome. HIV-1 primarily infects CD4 T lymphocytes and macrophages using the CD4 molecule on these cells as the primary receptor and CCR5/CXCR4 chemokine receptors as co-receptors (Loetscher *et al.*, 2000). During the early phase of the viral life cycle the HIV-1 DNA integrates into the host genome: latency at this stage is called pre-integration latency referring to the generation of different forms of viral DNA before integration. Generation of complete virions after viral replication occurs at the late phase. Post-integration latency at this stage refers to the lack of replication after the insertion of viral DNA into the host genome (Stevenson, 2003, Lassen *et al.*, 2004). Many different mechanisms have been proposed to establish and maintain this post-integration latency in HIV-1 infection. Some of the more important ones are thought to be: viral integration sites, chromatin environment, lack of key transcription factors, impaired viral activator Tat and RNA interference (Dahl *et al.*, Le Douce *et al.*, Alexaki *et al.*, 2008, Marcello, 2006, Margolis). Natural reversion of an infected T cell to quiescent state, as with memory cells, results in viral latency in these resting cells. Lack of host factors like NF-kB and transcription elongation factor pTEFb restricts HIV transcription in resting cells leading to establishment of latency (Williams and Greene, 2007). Macrophages and dendritic cells act as viral reservoirs as they are less susceptible to cell death as compared to CD4+ T cells (Le Douce *et al.*). Viral replication in macrophages is characterized by longer persistence of unintegrated DNA(Kelly *et al.*, 2008, Swingler *et al.*, 1999), production of chemokines and soluble mediators (Swingler *et al.*, 1999) and generation of infective viral particles (Sharova *et al.*, 2005). Dendritic cells are thought to capture viruses using the DC-SIGN (DC-specific ICAM3-grabbing non-integrin) surface receptor and efficiently transmitting to T cells *via* the immunological synapse during antigen presentation (Alexaki *et al.*, 2008). Follicular DCs in lymphoid tissues are specialized in trapping virus particles in the form of immune complexes and may provide for viral persistence (Smith-Franklin *et al.*, 2002, Keele *et al.*, 2008). Anatomical compartments (spleen, lymph nodes and gut-associated lymphoid tissue (GALT)) where HIV replication can take place in the presence of HAART (highly active anti-retroviral therapy) due to poor penetration of anti-retrovirals or due to special biological properties of the compartments such as being in an immunoprivileged sites (CNS) are also sites of viral persistence (Gunthard *et al.*, 2001, Dahl *et al.*).

CONFLICT OF INTEREST

The authors confirm that this chapter contents have no conflict of interest.

ACKNOWLEDGEMENTS

Declared None.

REFERENCES

ABRAMSON, A. L., NOURI, M., MULLOOLY, V., FISCH, G. & STEINBERG, B. M. 2004. Latent Human Papillomavirus infection is comparable in the larynx and trachea. *J Med Virol,* 72, 473-7.
ALEXAKI, A., LIU, Y. & WIGDAHL, B. 2008. Cellular reservoirs of HIV-1 and their role in viral persistence. *Curr HIV Res,* 6, 388-400.
BABCOCK, G. J. & THORLEY-LAWSON, D. A. 2000. Tonsillar memory B cells, latently infected with Epstein-Barr virus, express the restricted pattern of latent genes previously found only in Epstein-Barr virus-associated tumors. *Proc Natl Acad Sci U S A,* 97, 12250-5.
BAIRD, N. L., YU, X., COHRS, R. J. & GILDEN, D. 2013. Varicella zoster virus (VZV)-human neuron interaction. *Viruses,* 5, 2106-15.
BEAUDENON, S., KREMSDORF, D., CROISSANT, O., JABLONSKA, S., WAIN-HOBSON, S. & ORTH, G. 1986. A novel type of human papillomavirus associated with genital neoplasias. *Nature,* 321, 246-9.
BEGO, M., MACIEJEWSKI, J., KHAIBOULLINA, S., PARI, G. & ST JEOR, S. 2005. Characterization of an antisense transcript spanning the UL81-82 locus of human cytomegalovirus. *J Virol,* 79, 11022-34.
BENTZ, G. L., MOSS, C. R., 2ND, WHITEHURST, C. B., MOODY, C. A. & PAGANO, J. S. 2015. LMP1-induced Sumoylation Influences the Maintenance of EBV Latency Through KAP1. *J Virol.*
BILLAUD, G., THOUVENOT, D. & MORFIN, F. 2009. Drug targets in herpes simplex and Epstein Barr Virus infections. *Infect Disord Drug Targets,* 9, 117-25.

BODAGHI, B., JONES, T. R., ZIPETO, D., VITA, C., SUN, L., LAURENT, L., ARENZANA-SEISDEDOS, F., VIRELIZIER, J. L. & MICHELSON, S. 1998. Chemokine sequestration by viral chemoreceptors as a novel viral escape strategy: withdrawal of chemokines from the environment of cytomegalovirus-infected cells. *J Exp Med,* 188, 855-66.

CASTILLO, J. P. & KOWALIK, T. F. 2002. Human cytomegalovirus immediate early proteins and cell growth control. *Gene,* 290, 19-34.

COHRS, R. J. & GILDEN, D. H. 2003. Varicella zoster virus transcription in latently-infected human ganglia. *Anticancer Res,* 23, 2063-9.

COLBERG-POLEY, A. M. 1996. Functional roles of immediate early proteins encoded by the human cytomegalovirus UL36-38, UL115-119, TRS1/IRS1 and US3 loci. *Intervirology,* 39, 350-60.

CROUGH, T. & KHANNA, R. 2009. Immunobiology of human cytomegalovirus: from bench to bedside. *Clin Microbiol Rev,* 22, 76-98, Table of Contents.

CUNNINGHAM, A. L., DIEFENBACH, R. J., MIRANDA-SAKSENA, M., BOSNJAK, L., KIM, M., JONES, C. & DOUGLAS, M. W. 2006. The cycle of human herpes simplex virus infection: virus transport and immune control. *J Infect Dis,* 194 Suppl 1, S11-8.

D'SA-EIPPER, C., SUBRAMANIAN, T. & CHINNADURAI, G. 1996. bfl-1, a bcl-2 homologue, suppresses p53-induced apoptosis and exhibits potent cooperative transforming activity. *Cancer Res,* 56, 3879-82.

DAHL, V., JOSEFSSON, L. & PALMER, S. HIV reservoirs, latency, and reactivation: prospects for eradication. *Antiviral Res,* 85, 286-94.

DAVISON, A. J. & SCOTT, J. E. 1986. The complete DNA sequence of varicella-zoster virus. *J Gen Virol,* 67 (Pt 9), 1759-816.

DE VILLIERS, E. M., FAUQUET, C., BROKER, T. R., BERNARD, H. U. & ZUR HAUSEN, H. 2004. Classification of papillomaviruses. *Virology,* 324, 17-27.

DONAHUE, J. G., CHOO, P. W., MANSON, J. E. & PLATT, R. 1995. The incidence of herpes zoster. *Arch Intern Med,* 155, 1605-9.

DOORBAR, J. 2006. Molecular biology of human papillomavirus infection and cervical cancer. *Clin Sci (Lond),* 110, 525-41.

EFSTATHIOU, S. & PRESTON, C. M. 2005. Towards an understanding of the molecular basis of herpes simplex virus latency. *Virus Res,* 111, 108-19.

EISFELD, A. J., TURSE, S. E., JACKSON, S. A., LERNER, E. C. & KINCHINGTON, P. R. 2006. Phosphorylation of the varicella-zoster virus (VZV) major transcriptional regulatory protein IE62 by the VZV open reading frame 66 protein kinase. *J Virol,* 80, 1710-23.

FINGER, R., HUGHES, J. P., MEADE, B. J., PELLETIER, A. R. & PALMER, C. T. 1994. Age-specific incidence of chickenpox. *Public Health Rep,* 109, 750-5.

FIORINA, L., RICOTTI, M., VANOLI, A., LUINETTI, O., DALLERA, E., RIBONI, R., PAOLUCCI, S., BRUGNATELLI, S., PAULLI, M., PEDRAZZOLI, P., BALDANTI, F. & PERFETTI, V. 2014. Systematic analysis of human oncogenic viruses in colon cancer revealed EBV latency in lymphoid infiltrates. *Infect Agent Cancer,* 9, 18.

GILDEN, D. H., ROZENMAN, Y., MURRAY, R., DEVLIN, M. & VAFAI, A. 1987. Detection of varicella-zoster virus nucleic acid in neurons of normal human thoracic ganglia. *Ann Neurol,* 22, 377-80.

GISSMANN, L., DIEHL, V., SCHULTZ-COULON, H. J. & ZUR HAUSEN, H. 1982. Molecular cloning and characterization of human papilloma virus DNA derived from a laryngeal papilloma. *J Virol,* 44, 393-400.

GOODRUM, F. D., JORDAN, C. T., HIGH, K. & SHENK, T. 2002. Human cytomegalovirus gene expression during infection of primary hematopoietic progenitor cells: a model for latency. *Proc Natl Acad Sci U S A,* 99, 16255-60.

GUNTHARD, H. F., HAVLIR, D. V., FISCUS, S., ZHANG, Z. Q., ERON, J., MELLORS, J., GULICK, R., FROST, S. D., BROWN, A. J., SCHLEIF, W., VALENTINE, F., JONAS, L., MEIBOHM, A., IGNACIO, C. C., ISAACS, R., GAMAGAMI, R., EMINI, E., HAASE, A., RICHMAN, D. D. & WONG, J. K. 2001. Residual human immunodeficiency virus (HIV) Type 1 RNA and DNA in lymph nodes and HIV RNA in genital secretions and in cerebrospinal fluid after suppression of viremia for 2 years. *J Infect Dis,* 183, 1318-27.

HENDERSON, G., PENG, W., JIN, L., PERNG, G. C., NESBURN, A. B., WECHSLER, S. L. & JONES, C. 2002. Regulation of caspase 8- and caspase 9-induced apoptosis by the herpes simplex virus type 1 latency-associated transcript. *J Neurovirol,* 8 Suppl 2, 103-11.

HOAGLAND, R. J. 1955. The transmission of infectious mononucleosis. *Am J Med Sci,* 229, 262-72.

HOLLEY-GUTHRIE, E. A., QUINLIVAN, E. B., MAR, E. C. & KENNEY, S. 1990. The Epstein-Barr virus (EBV) BMRF1 promoter for early antigen (EA-D) is regulated by the EBV transactivators, BRLF1 and BZLF1, in a cell-specific manner. *J Virol,* 64, 3753-9.

JONES, C. 2003. Herpes simplex virus type 1 and bovine herpesvirus 1 latency. *Clin Microbiol Rev,* 16, 79-95.

JOSEPH, A. M., BABCOCK, G. J. & THORLEY-LAWSON, D. A. 2000. EBV persistence involves strict selection of latently infected B cells. *J Immunol,* 165, 2975-81.

JUNG, W. W., CHUN, T., SUL, D., HWANG, K. W., KANG, H. S., LEE, D. J. & HAN, I. K. 2004. Strategies against human papillomavirus infection and cervical cancer. *J Microbiol,* 42, 255-66.

KALEJTA, R. F. 2008a. Functions of human cytomegalovirus tegument proteins prior to immediate early gene expression. *Curr Top Microbiol Immunol,* 325, 101-15.

KALEJTA, R. F. 2008b. Tegument proteins of human cytomegalovirus. *Microbiol Mol Biol Rev,* 72, 249-65, table of contents.

KANE, M. & GOLOVKINA, T. 2010. Common threads in persistent viral infections. *J Virol,* 84, 4116-23.

KEELE, B. F., TAZI, L., GARTNER, S., LIU, Y., BURGON, T. B., ESTES, J. D., THACKER, T. C., CRANDALL, K. A., MCARTHUR, J. C. & BURTON, G. F. 2008. Characterization of the follicular dendritic cell reservoir of human immunodeficiency virus type 1. *J Virol,* 82, 5548-61.

KELLY, J., BEDDALL, M. H., YU, D., IYER, S. R., MARSH, J. W. & WU, Y. 2008. Human macrophages support persistent transcription from unintegrated HIV-1 DNA. *Virology,* 372, 300-12.

KENNEDY, P. G. & COHRS, R. J. 2010. Varicella-zoster virus human ganglionic latency: a current summary. *J Neurovirol,* 16, 411-8.

KENNEY, S., HOLLEY-GUTHRIE, E., MAR, E. C. & SMITH, M. 1989a. The Epstein-Barr virus BMLF1 promoter contains an enhancer element that is responsive to the BZLF1 and BRLF1 transactivators. *J Virol,* 63, 3878-83.

KENNEY, S., KAMINE, J., HOLLEY-GUTHRIE, E., LIN, J. C., MAR, E. C. & PAGANO, J. 1989b. The Epstein-Barr virus (EBV) BZLF1 immediate-early gene product differentially affects latent *versus* productive EBV promoters. *J Virol,* 63, 1729-36.

KENT, J. R., KANG, W., MILLER, C. G. & FRASER, N. W. 2003. Herpes simplex virus latency-associated transcript gene function. *J Neurovirol,* 9, 285-90.

KILGORE, P. E., KRUSZON-MORAN, D., SEWARD, J. F., JUMAAN, A., VAN LOON, F. P., FORGHANI, B., MCQUILLAN, G. M., WHARTON, M., FEHRS, L. J., COSSEN, C. K. & HADLER, S. C. 2003. Varicella in Americans from NHANES III: implications for control through routine immunization. *J Med Virol,* 70 Suppl 1, S111-8.

KUDESIA, G., PARTRIDGE, S., FARRINGTON, C. P. & SOLTANPOOR, N. 2002. Changes in age related seroprevalence of antibody to varicella zoster virus: impact on vaccine strategy. *J Clin Pathol,* 55, 154-5.

KUTOK, J. L. & WANG, F. 2006. Spectrum of Epstein-Barr virus-associated diseases. *Annu Rev Pathol,* 1, 375-404.

LAGUARDIA, J. J., COHRS, R. J. & GILDEN, D. H. 1999. Prevalence of varicella-zoster virus DNA in dissociated human trigeminal ganglion neurons and nonneuronal cells. *J Virol,* 73, 8571-7.

LAI, K. Y., CHOU, Y. C., LIN, J. H., LIU, Y., LIN, K. M., DOONG, S. L., CHEN, M. R., YEH, T. H., LIN, S. J. & TSAI, C. H. 2015. Maintenance of Epstein-Barr Virus Latent Status by a Novel Mechanism, Latent Membrane Protein 1-Induced Interleukin-32, *via* the Protein Kinase Cdelta Pathway. *J Virol,* 89, 5968-80.

LANDOLFO, S., GARIGLIO, M., GRIBAUDO, G. & LEMBO, D. 2003. The human cytomegalovirus. *Pharmacol Ther,* 98, 269-97.

LANFRANCA, M. P., MOSTAFA, H. H. & DAVIDO, D. J. 2014. HSV-1 ICP0: An E3 Ubiquitin Ligase That Counteracts Host Intrinsic and Innate Immunity. *Cells,* 3, 438-54.

LARSSON, S., SODERBERG-NAUCLER, C. & MOLLER, E. 1998a. Productive cytomegalovirus (CMV) infection exclusively in CD13-positive peripheral blood mononuclear cells from CMV-infected individuals: implications for prevention of CMV transmission. *Transplantation,* 65, 411-5.

LARSSON, S., SODERBERG-NAUCLER, C., WANG, F. Z. & MOLLER, E. 1998b. Cytomegalovirus DNA can be detected in peripheral blood mononuclear cells from all seropositive and most seronegative healthy blood donors over time. *Transfusion,* 38, 271-8.

LASSEN, K., HAN, Y., ZHOU, Y., SILICIANO, J. & SILICIANO, R. F. 2004. The multifactorial nature of HIV-1 latency. *Trends Mol Med,* 10, 525-31.

LE DOUCE, V., HERBEIN, G., ROHR, O. & SCHWARTZ, C. Molecular mechanisms of HIV-1 persistence in the monocyte-macrophage lineage. *Retrovirology,* 7, 32.

LIU, X. F., WANG, X., YAN, S., ZHANG, Z., ABECASSIS, M. & HUMMEL, M. 2013. Epigenetic control of cytomegalovirus latency and reactivation. *Viruses,* 5, 1325-45.

LOETSCHER, P., MOSER, B. & BAGGIOLINI, M. 2000. Chemokines and their receptors in lymphocyte traffic and HIV infection. *Adv Immunol,* 74, 127-80.

LOOKER, K. J., GARNETT, G. P. & SCHMID, G. P. 2008. An estimate of the global prevalence and incidence of herpes simplex virus type 2 infection. *Bull World Health Organ,* 86, 805-12, A.

MACSWEEN, K. F. & CRAWFORD, D. H. 2003. Epstein-Barr virus-recent advances. *Lancet Infect Dis,* 3, 131-40.

MAHALINGAM, R., WELLISH, M., WOLF, W., DUELAND, A. N., COHRS, R., VAFAI, A. & GILDEN, D. 1990. Latent varicella-zoster viral DNA in human trigeminal and thoracic ganglia. *N Engl J Med,* 323, 627-31.

MAHALINGAM, R., WELLISH, M. C., DUELAND, A. N., COHRS, R. J. & GILDEN, D. H. 1992. Localization of herpes simplex virus and varicella zoster virus DNA in human ganglia. *Ann Neurol,* 31, 444-8.

MARAN, A., AMELLA, C. A., DI LORENZO, T. P., AUBORN, K. J., TAICHMAN, L. B. & STEINBERG, B. M. 1995. Human papillomavirus type 11 transcripts are present at low abundance in latently infected respiratory tissues. *Virology,* 212, 285-94.

MARCELLO, A. 2006. Latency: the hidden HIV-1 challenge. *Retrovirology,* 3, 7.

MARGOLIS, D. M. Mechanisms of HIV latency: an emerging picture of complexity. *Curr HIV/AIDS Rep,* 7, 37-43.

MCCORMICK, A. L. & MOCARSKI, E. S. 2015. The immunological underpinnings of vaccinations to prevent cytomegalovirus disease. *Cell Mol Immunol,* 12, 170-9.

MENDELSON, M., MONARD, S., SISSONS, P. & SINCLAIR, J. 1996. Detection of endogenous human cytomegalovirus in CD34+ bone marrow progenitors. *J Gen Virol,* 77 (Pt 12), 3099-102.

MORIUCHI, H., MORIUCHI, M. & COHEN, J. I. 1995. The varicella-zoster virus immediate-early 62 promoter contains a negative regulatory element that binds transcriptional factor NF-Y. *Virology,* 214, 256-8.

MORIUCHI, M., MORIUCHI, H., STRAUS, S. E. & COHEN, J. I. 1994. Varicella-zoster virus (VZV) virion-associated transactivator open reading frame 62 protein enhances the infectivity of VZV DNA. *Virology,* 200, 297-300.

MOUNTS, P., SHAH, K. V. & KASHIMA, H. 1982. Viral etiology of juvenile- and adult-onset squamous papilloma of the larynx. *Proc Natl Acad Sci U S A,* 79, 5425-9.

NAGEL, M. A. & GILDEN, D. 2014. Neurological complications of varicella zoster virus reactivation. *Curr Opin Neurol,* 27, 356-60.

NAHMIAS, A. J. & ROIZMAN, B. 1973. Infection with herpes-simplex viruses 1 and 2. 3. *N Engl J Med,* 289, 781-9.

NICHOLSON, I. P., SUTHERLAND, J. S., CHAUDRY, T. N., BLEWETT, E. L., BARRY, P. A., NICHOLL, M. J. & PRESTON, C. M. 2009. Properties of virion transactivator proteins encoded by primate cytomegaloviruses. *Virol J,* 6, 65.

OUDEJANS, J. J., VAN DEN BRULE, A. J., JIWA, N. M., DE BRUIN, P. C., OSSENKOPPELE, G. J., VAN DER VALK, P., WALBOOMERS, J. M. & MEIJER, C. J. 1995. BHRF1, the Epstein-Barr virus (EBV) homologue of the BCL-2 protooncogene, is transcribed in EBV-associated B-cell lymphomas and in reactive lymphocytes. *Blood,* 86, 1893-902.

PANG, M. F., LIN, K. W. & PEH, S. C. 2009. The signaling pathways of Epstein-Barr virus-encoded latent membrane protein 2A (LMP2A) in latency and cancer. *Cell Mol Biol Lett,* 14, 222-47.

PERNG, G. C., JONES, C., CIACCI-ZANELLA, J., STONE, M., HENDERSON, G., YUKHT, A., SLANINA, S. M., HOFMAN, F. M., GHIASI, H., NESBURN, A. B. & WECHSLER, S. L. 2000.

Virus-induced neuronal apoptosis blocked by the herpes simplex virus latency-associated transcript. *Science*, 287, 1500-3.

ROIZMAN, B. & WHITLEY, R. J. 2013. An inquiry into the molecular basis of HSV latency and reactivation. *Annu Rev Microbiol*, 67, 355-74.

SCHLEISS, M. R., BUUS, R., CHOI, K. Y. & MCGREGOR, A. 2013. An Attenuated CMV Vaccine with a Deletion in Tegument Protein GP83 (pp65 Homolog) Protects against Placental Infection and Improves Pregnancy Outcome in a Guinea Pig Challenge Model. *Future Virol*, 8, 1151-1160.

SCHNEIDER, A. 1993. Pathogenesis of genital HPV infection. *Genitourin Med*, 69, 165-73.

SCHOLZ, M., DOERR, H. W. & CINATL, J. 2003. Human cytomegalovirus retinitis: pathogenicity, immune evasion and persistence. *Trends Microbiol*, 11, 171-8.

SEDMAK, D. D., GUGLIELMO, A. M., KNIGHT, D. A., BIRMINGHAM, D. J., HUANG, E. H. & WALDMAN, W. J. 1994. Cytomegalovirus inhibits major histocompatibility class II expression on infected endothelial cells. *Am J Pathol*, 144, 683-92.

SHAROVA, N., SWINGLER, C., SHARKEY, M. & STEVENSON, M. 2005. Macrophages archive HIV-1 virions for dissemination in trans. *EMBO J*, 24, 2481-9.

SINCLAIR, J. & SISSONS, P. 2006. Latency and reactivation of human cytomegalovirus. *J Gen Virol*, 87, 1763-79.

SINCLAIR, J. H., BAILLIE, J., BRYANT, L. A., TAYLOR-WIEDEMAN, J. A. & SISSONS, J. G. 1992. Repression of human cytomegalovirus major immediate early gene expression in a monocytic cell line. *J Gen Virol*, 73 (Pt 2), 433-5.

SISSONS, J. G., CARMICHAEL, A. J., MCKINNEY, N., SINCLAIR, J. H. & WILLS, M. R. 2002. Human cytomegalovirus and immunopathology. *Springer Semin Immunopathol*, 24, 169-85.

SMITH-FRANKLIN, B. A., KEELE, B. F., TEW, J. G., GARTNER, S., SZAKAL, A. K., ESTES, J. D., THACKER, T. C. & BURTON, G. F. 2002. Follicular dendritic cells and the persistence of HIV infectivity: the role of antibodies and Fcgamma receptors. *J Immunol*, 168, 2408-14.

STEVENS, J. G., WAGNER, E. K., DEVI-RAO, G. B., COOK, M. L. & FELDMAN, L. T. 1987. RNA complementary to a herpesvirus alpha gene mRNA is prominent in latently infected neurons. *Science*, 235, 1056-9.

STEVENSON, M. 2003. HIV-1 pathogenesis. *Nat Med*, 9, 853-60.

STEWART, J. P. & ROONEY, C. M. 1992. The interleukin-10 homolog encoded by Epstein-Barr virus enhances the reactivation of virus-specific cytotoxic T cell and HLA-unrestricted killer cell responses. *Virology*, 191, 773-82.

STUART, A. D., STEWART, J. P., ARRAND, J. R. & MACKETT, M. 1995. The Epstein-Barr virus encoded cytokine viral interleukin-10 enhances transformation of human B lymphocytes. *Oncogene*, 11, 1711-9.

STUBENRAUCH, F. & LAIMINS, L. A. 1999. Human papillomavirus life cycle: active and latent phases. *Semin Cancer Biol*, 9, 379-86.

SWINGLER, S., MANN, A., JACQUE, J., BRICHACEK, B., SASSEVILLE, V. G., WILLIAMS, K., LACKNER, A. A., JANOFF, E. N., WANG, R., FISHER, D. & STEVENSON, M. 1999. HIV-1 Nef mediates lymphocyte chemotaxis and activation by infected macrophages. *Nat Med*, 5, 997-103.

TAYLOR-WIEDEMAN, J., SISSONS, P. & SINCLAIR, J. 1994. Induction of endogenous human cytomegalovirus gene expression after differentiation of monocytes from healthy carriers. *J Virol*, 68, 1597-604.

THOMPSON, M. P. & KURZROCK, R. 2004. Epstein-Barr virus and cancer. *Clin Cancer Res*, 10, 803-21.

THOMPSON, R. L. & SAWTELL, N. M. 2000. Replication of herpes simplex virus type 1 within trigeminal ganglia is required for high frequency but not high viral genome copy number latency. *J Virol*, 74, 965-74.

THORLEY-LAWSON, D. A. 2001. Epstein-Barr virus: exploiting the immune system. *Nat Rev Immunol*, 1, 75-82.

THORLEY-LAWSON, D. A. 2005. EBV the prototypical human tumor virus--just how bad is it? *J Allergy Clin Immunol*, 116, 251-61; quiz 262.

TSURUMI, T., FUJITA, M. & KUDOH, A. 2005. Latent and lytic Epstein-Barr virus replication strategies. *Rev Med Virol*, 15, 3-15.

VANCIKOVA, Z. & DVORAK, P. 2001. Cytomegalovirus infection in immunocompetent and immunocompromised individuals--a review. *Curr Drug Targets Immune Endocr Metabol Disord,* 1, 179-87.

VARANI, S., FRASCAROLI, G., GIBELLINI, D., POTENA, L., LAZZAROTTO, T., LEMOLI, R. M., MAGELLI, C., SODERBERG-NAUCLER, C. & LANDINI, M. P. 2005. Impaired dendritic cell immunophenotype and function in heart transplant patients undergoing active cytomegalovirus infection. *Transplantation,* 79, 219-27.

WATSON, Z., DHUMMAKUPT, A., MESSER, H., PHELAN, D. & BLOOM, D. 2013. Role of polycomb proteins in regulating HSV-1 latency. *Viruses,* 5, 1740-57.

WHITLEY, R. J. 2002. Herpes simplex virus infection. *Semin Pediatr Infect Dis,* 13, 6-11.

WILLIAMS, S. A. & GREENE, W. C. 2007. Regulation of HIV-1 latency by T-cell activation. *Cytokine,* 39, 63-74.

YOUNG, L. S. & MURRAY, P. G. 2003. Epstein-Barr virus and oncogenesis: from latent genes to tumours. *Oncogene,* 22, 5108-21.

YOUNG, L. S. & RICKINSON, A. B. 2004. Epstein-Barr virus: 40 years on. *Nat Rev Cancer,* 4, 757-68.

YUROCHKO, A. D. 2008. Human cytomegalovirus modulation of signal transduction. *Curr Top Microbiol Immunol,* 325, 205-20.

ZERBONI, L. & ARVIN, A. 2011. Investigation of varicella-zoster virus neurotropism and neurovirulence using SCID mouse-human DRG xenografts. *J Neurovirol,* 17, 570-7.

ZUR HAUSEN, H. 2009. Papillomaviruses in the causation of human cancers - a brief historical account. *Virology,* 384, 260-5.

CHAPTER 3

Herpes Simplex Virus Infections and Vaccine Advances

Tu Thanh Mai[1] and Liljana Stevceva[2,*]

[1]The Paul L. Foster School of Medicine, El Paso, TX, USA and [2]California Nothstate University College of Medicine' USA

Abstract: Herpes simplex viruses belong to the subfamily *Alphaherpesvirinae* and the genus Simplexvirus and have the capacity to establish latent infections in sensory ganglia of humans. Intermittent reactivation of the virus can cause retrograde transportation to the dermatome where initial infection occurred, the virus crosses into the stratified squamous epithelium where it starts replicating again. Herpes simplex viruses that cause infections in humans are HSV-1 (infection of the skin and oral mucosa) and HSV-2 (sexually transmitted infection of the genital tract). Despite almost a century old efforts to develop vaccine against HSV viral infection, results from clinical trials in humans have been disappointing. Three main approaches were used to develop the vaccine: glycoprotein vaccine (gB and gD), mutated inactivated live virus and DNA vaccines. Although all vaccine candidates showed excellent results in animal models, significant protection was not seen in human clinical trials. This not only questions the validity of the animal models but also calls for a change in the current strategies in designing future vaccine candidates.

In designing the vaccine candidates very little attention was devoted to the immune evasion mechanisms of the virus that are the main reason that the virus is able to persist and to reactivate when immune responses weaken.

Keywords: Herpes, alphaherpesviridae, sensory ganglia, HSV-1, HSV-2, glycoprotein, vaccine, immune evasion, latent infection, skin infection, genital tract infection, viral envelope, DNA vaccine, STD, Simplexvirus, gB, gD, CD8+ T cells exhaustion, HERPEVAC, ICP4.

INTRODUCTION

Herpes simplex virus 1 and 2 (HSV-1 and HSV-2) are members of the family *Herpesviridae.* They belong to the subfamily *Alphaherpesvirinae* and the genus *Simplexvirus.* Characteristics of the subfamily *Alphaherpesvirinae* include a

---

*Corresponding author Liljana Stevceva: University of Texas Rio Grande Valley School of Medicine, 2102 Treasure Hills Blvd., Harlingen TX, USA; E-mail: Liljana@hotmail.com

relative short reproductive cycle, efficient destruction of infected cells with release of viral progeny, rapid spread in culture, and capacity to establish latent infection in sensory ganglia (Long, 2009).

Structure of the Virus

Virion particles are approximately 110 to 120nm in diameter (Long, 2009). As other members of the family *Herpesviridae*, HSV-1 and HSV-2 share the same virion structure with four layers: core, capsid, tegument, and envelope (Fig. **1a**). A linear double-stranded DNA genome, contained within a central core, is surrounded by an icosahedral capsid. The capsid is surrounded by tegument, a tightly adherent membrane consisting of viral proteins. An envelope, a lipid bilayer derived from host cell membranes, loosely surrounds the icosahedral capsid and tegument (Remington, 2006). Despite many similar characteristics in DNA homology, antigenic determinants, tissue tropism, and disease symptoms, HSV-1 and HSV-2 can still be distinguished by subtle but considerable differences in these identities (Murray, 2008).

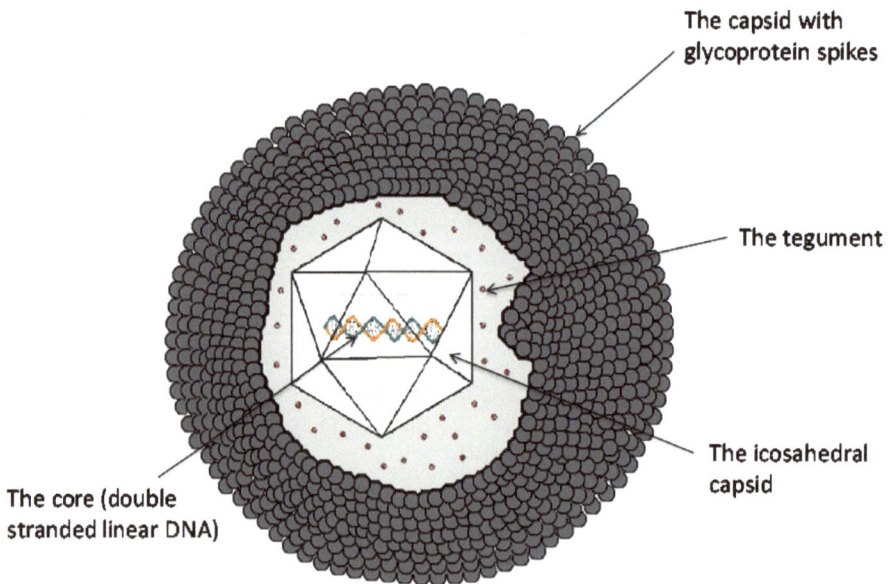

Fig. (1a). HSV particle contains four layers: envelope, tegument, capsid, and core.

The Envelope

Besides lipids and polyamines, the envelope consists of 600 to 750 viral glycoprotein spikes that vary in length, spacing and in the angles at which they emerge from the membrane. Their distribution on the envelope is nonrandom, suggesting functional clustering (Grunewald *et al.*, 2003). At least 11 virally encoded glycoproteins have been identified, including B, C, D, E, G, H, I, J, K, L and M (Ruocco *et al.*, 2007). Although HSV-1 and HSV-2 share similarities in their DNA sequence, their envelope glycoproteins provide them distinctive properties and unique antigens to which the host is responding (Fatahzadeh and Schwartz, 2007).

Among the 11 viral glycoproteins, only gB, gD, gH, and gL (Fig. **1b**) are required for cell entry (Reske *et al.*, 2007).

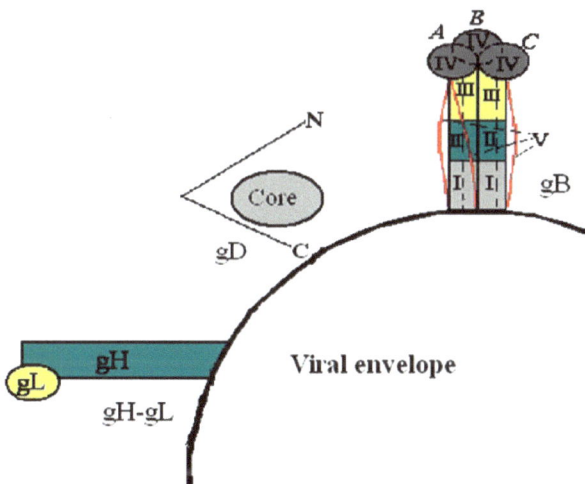

Fig. (1b). Four envelope glycoproteins imporant for cell entry (gB, gD, gH, and gL). GB is a trimer composed of 3 protomers A, B, and C. Each protomer has five domains (I, II, III, IV, and V). gD core has a V-like shape with C-terminal attaching to the viral membrane. gH-gL functions together as a heterodimer.

HSV-1 gB is the most conserved glycoprotein structure, resembling the structure of vesicular stomatitis virus glycoportein G (VSV G). It contains 904 residues and is composed of three protomers A, B, and C. Each protomer has five distinct domains. These domains include: I (base), locating proximal to the viral membrane; II (middle); III (core); IV (crown); and V (arm), a long extension from top to bottom of the molecule. Trimeric gB spike has the approximate dimensions of $85A^0$ by $80A^0$ by $160A^0$ (Heldwein *et al.*, 2006).

HSV-1 gD contains 369 residues with an N terminal ectodomain composed of 316 residues and three N-linked oligosaccharide attachment sites (Watson and Vande

Woude, 1982). The core of gD comprises of a V-like immunoglobulin fold, flanked by large N- and C- terminal extensions. N-terminal extension has a hairpin structure and comprises of all the HveA (Herpes virus entry mediator A) binding sites. C-terminal extension wraps around gD core and anchors the glycoprotein into the viral membrane (Carfi *et al*., 2001). Studies in HSV-infected individuals showed that majority of the neutralizing antibody responses are directed against gD or gD + gB. However, the cohort of HSV-infected individuals with high recurrence rates had higher titer of neutralizing antibodies and, although levels of neutralizing antibodies correlated with the severity of the infection, they did not confer protection from recurrence (Cairns *et al*., 2015).

HSV-1 gH and gL function together as a heterodimer (Hutchinson *et al*., 1992). HSV-1 gH is an 838-residue glycoprotein with 7 N-glycosylation sites and 8 cysteine residues in the ectodomain region (Cairns *et al*., 2005). Whereas the C-terminal region of gH is needed for fusion with the viral membrane, the N-terminal portion is needed for interaction with gL (Cairns *et al*., 2005), (Cairns *et al*., 2003). HSV-1 gL contains 224 amino acids and is associated with the viral envelope upon complexing with gH (Peng *et al*., 1998), (Dubin and Jiang, 1995). GH is required for the expression of gL on the cell surface, and only a small fraction of gL is processed in the absence of gH (Hutchinson *et al*., 1992).

HSV Icosahedral Capsid

The capsid functions to protect the viral DNA from external damages and to release the viral DNA into the host cell nucleus after a new cycle of infection (Newcomb *et al*., 2003). Cryo-electron tomography has shown that the capsid occupies approximately one-third of the volume enclosed within the envelope, whereas the tegument occupies the remaining two-thirds. It was also shown not to be positioned at the geometric center of the virion particle (Grunewald *et al*., 2003).

The capsid of HSV-1 is 125nm in diameter and 15nm thick. Its major structural features are 162 capsomers, comprising of 150 hexons and 12 pentons (Fig. **1c**). 320 triplexes (trivalent structures composed of two copies of UL18 and one copies of UL38) connect the capsomers in groups of three. The pentons are located at the capsid vertices, whereas the hexons are located at the capsid edges and faces (Homa and Brown, 1997), (Scholtes and Baines, 2009). Each of the 150 hexons is composed of 6 copies of the major capsid protein VP5 and six copies of VP26 (Newcomb *et al*., 2003). Eleven of the twelve pentons each are composed of 5 copies of the VP5; whereas the twelfth penton is composed of twelve copies of the UL6 genes, and serves as the portal through which DNA is inserted (Scholtes and Baines, 2009).

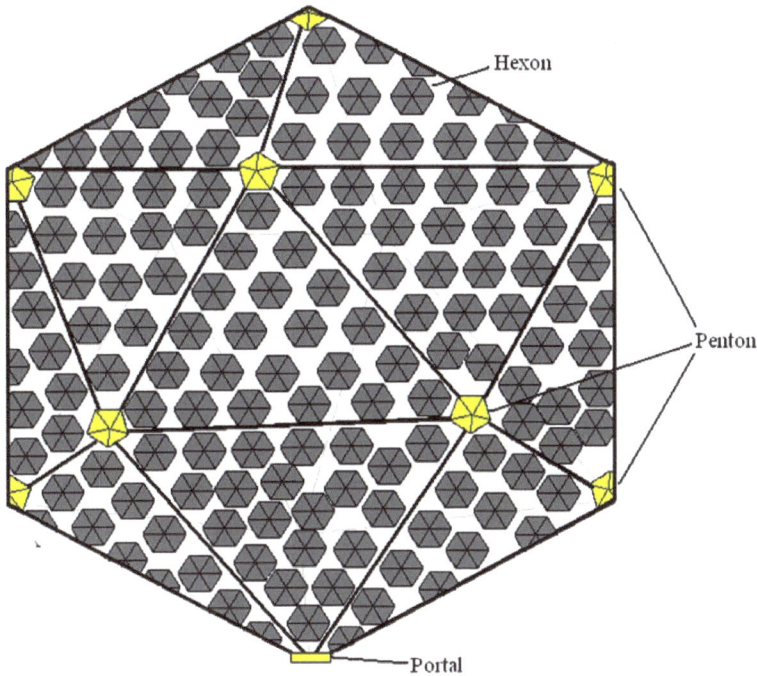

Fig. (1c). HSV-1 icosahedral capsid consists of 162 capsomeres, including 150 hexons locating at faces and edges, and 11 pentons and one portal locating at 12 vertices.

Tegument

The tegument is the space between the nucleocapsid surface and the inner surface of the virion envelope. Its structure is the least well characterized compared to other virion compartments (Zhou *et al.*, 1999). A study using cryo-electron tomogram has shown that the tegument has a well-defined size and shape and forms an asymmetric cap around the nucleocapsid. This same study has also shown that the tegument substructure is particulate and contains some short actin-like filaments (Grunewald *et al.*, 2003).

The tegument contains approximately 20 different viral proteins, and some of those are essential for the virus. Examples include the capsid transport protein (UL36), the viral transcription initiating protein (ICP0, ICP4, and UL48), and the viral translation regulating protein (UL41) (Kelly *et al.*, 2009). Tegument proteins present in large numbers (1000 to 2000 copies per virion) tend to play a structural role (such as UL46, UL47, UL48, and UL49), whereas those present in small amounts (100 to 150 copies per virion) tend to perform regulatory roles (such as ICP0 and ICP4) (Vittone *et al.*, 2005).

The genome of HSV consists of double-stranded linear DNA, with the approximate molecular weight of 100×10^6 (Mandell, 2005). At least 84 polypeptides are encoded in the HSV genome, which include: (1) viral envelope glycoproteins, (2) capsid proteins, (3) enzymes (DNA polymerase, protein kinase), (4) DNA-binding proteins involved in replication, and (5) other proteins with poorly understood functions (Whitley *et al.*, 1998).

The overall genome homology between HSV-1 and HSV-2 is approximately 50% (Baringer, 2008). The genome of HSV-1 composes of 107,943bp of base composition 68.3% G+C, whereas the total genome length of HSV-2 was determined to be 154,746bp with a G+C content of 70.4%. 72 genes encoding 70 distinct proteins were recognized in the whole genome of HSV-1, whereas 74 genes were identified in the HSV-2 genome (Dolan *et al.*, 1998), (Dolan *et al.*, 1998).

HSV-1 genome consists of two unique regions, U_L (unique long), comprising 82% and U_S (unique short) comprising 18%, separated by the repeated regions *a*, *b*, and *c* (Fig. **1d**). Because of the recombination of the *a* sequences, HSV-1 genome exists as a mixture of four isomeric forms generated by inversion of the two types of unique sequences (U_L and U_S) relative to each other. Its genome contains three origins of DNA replication, also known as *cis*-acting elements. These origins of replication include one copy of Ori_L, located between UL29 and UL30 genes within U_L, and two copies of Ori_S, located within the repeated region *c* (Lehman and Boehmer, 1999).

Fig. (1d). HSV-1 genome contains two unique regions (U_L and U_S) separated by inverted repeats *a*, *b*, and *c* and three origins of replication (one Ori_L and two Ori_S).

Herpes viruses cause infections in humans characterized with appearance of painful blisters on the skin and mucosa of the mouth or the genital tract. Although

Herpes Virus 1 (HSV-1) predominantly causes infections of the lips and mouth and Herpes Virus 2 (HSV-2) of the genital tract both viruses can cause infection in either place (Bergelson, 2008). Infection with HSV-1 occurs *via* non-sexual transmission and is common with prevalence in the USA from 44% in 12-19 years olds to over 90% in people over 70 years of age. Infection with HSV-2 occurs *via* sexual transmission, is much less common and its prevalence in the USA is up to 28% (Smith and Robinson, 2002).

In most cases infection with herpes simplex viruses is asymptomatic. Sometimes, painful blisters occur near the site of viral entry. In a few days, the blisters dry out, crust and heal but can reappear later in life. Upon entering, the virus multiplies in the stratified squamous epithelium of the skin and then enters the unmyelinated sensory nerve fibers of the squamous epithelium from where it travels to the sensory ganglia. Here the virus establishes lifelong latency. Intermittent reactivation of the virus can cause retrograde transportation to the dermatome where initial infection occurred, the virus crosses into the stratified squamous epithelium where it starts replicating again. During reactivation, the virus is shed into the skin and mucosa and regardless whether clinical symptoms are present, the patient is infectious and can infect other people.

Both HSV-1 and HSV-2 cannot penetrate the stratum corneum of skin and initial infection occurs only where crack and abrasion exist that allow access of the virus to keratinocytes and Langerhans cells, cell types that express HSV receptors, nectin-1 and herpes virus entry mediator (Warner *et al.*, 1998), (Salio *et al.*, 1999).

Viral envelope is the most important part of the virus for infectivity. The envelope glycoproteins gB and gC bind to heparan sulfate on cell membrane. This binding is not sufficient for cell entry until binding of the glycoprotein gD to nectin 1, nectin 2 and HveM (Takai and Nakanishi, 2003), (Whitbeck *et al.*, 1997) triggers fusion of the viral envelope and the cell membrane after which the viral nucleocapsid and tegument enter the cell cytoplasm (Spear *et al.*, 2000). The virus expresses gC glycoprotein that binds C3b complement component and protects the virus from complement neutralization (reviewed by Friedman, (Friedman, 2003). The glycoprotein gE, also expressed on the viral envelope, binds the IgG Fc domain, blocking Fc-mediated activities, including complement activation and antibody-dependent cellular cytotoxicity. Mutant virus that lacks gC and gE glycoprotein is efficiently neutralized by complement even in the absence of neutralizing antibodies (McNearney *et al.*, 1987), (Hidaka *et al.*, 1991), (Friedman *et al.*, 1996), (Friedman, 2003).

Latency of the virus is maintained by CD8+ cells that are specific for the viral structural proteins (Khanna *et al.*, 2004). After primary infection, CD8+ T cells initially proliferate in the draining lymph nodes up to day 5, disseminate to lymphoid and non-lymphoid tissues where they differentiate into virus specific CD8 T cells after day 6 and peak around day 12, infiltrate the sensory ganglia and remain there. When CD8 a -/- mice were infected, the infection resulted in significantly higher mortality occurring after day 8. Thus, although CD8+ T cells seem to be occurring too late to prevent dissemination of the virus (virus is detected in trigeminal ganglia on day 2 and the brain on day 3), they do control the viral infection (Lang and Nikolich-Zugich, 2005). It has been shown that these cells also prevent the reactivation of the virus (Liu *et al.*, 2000). Secretion of IFNγ mediates, at least in part, that protection in the trigeminal ganglia (Liu *et al.*, 2001). In addition, the granule component granzyme B degrades the HSV-1 immediate early protein, ICP4, which is essential for further viral gene expression (Knickelbein *et al.*, 2008).

Constant exposure to antigen in the TG causes expression of PD-1 (programmed death cell receptor 1) that binds to PD-1 and PD-2 ligands and downregulates T cell function causing exhaustion. Tim 3 (T cell immunoglobulin and mucin domain-containing protein 3) is also upregulated and this has also been correlated with T cell exhaustion (Mott *et al.*, 2009). It was reported that CD8 T cell exhaustion occurs as a result of long-term (5 to 7 days), continuous exposure of virus-specific T cells to viral Ag and removal of Ag even for a short period (24 h) results in rapid recovery, decreased PD-1 expression, and lack of exhaustion (Keir *et al.*, 2008), (Mueller and Ahmed, 2009).

CD8+T cells exhaustion was related to the capacity of T cells to clear the virus and the status of latency in experiments done by Allen *et al.* (Allen *et al.*, 2011). The same report outlined the importance of LAT (latency-associated transcription) in HSV-1 reactivation and latency. The role of CD4+ T cells in HSV infection is poorly understood aside from the fact that CD4+ T cells help is required for the initial CD8+ T cell response but not for the secondary response (Jennings *et al.*, 1991).

CANDIDATE VACCINES AGAINST HSV INFECTIONS

Despite repeated attempts to develop a vaccine against herpes viruses, there is still no candidate vaccine that has been successful in affording protection. Most of vaccine development efforts so far have been focused on developing vaccine against HSV-2. Since many of the structural components of the two viruses are similar, vaccine based on one of the viruses should confer protection against the other. The following approaches have been tried:

Vaccine candidates based on envelope glycoproteins: Two of the envelope gycoproteins, gB and gD have been used as subunit vaccines. Chiron vaccine candidate consisting of recombinant forms of HSV-2 gB2 and gD2 lacking the carboxy-terminal regions combined with adjuvant MF59 was shown not to be protective in humans despite previously demonstrated efficacy in experimental models (Corey *et al.*, 1999). Recently published results from Phase 3 study GSK Herpevac vaccine candidate constructed from truncated glycoprotein D from HSV-2 strain G combined with ASO4 (alum and 3'-O deacylated-monophosphoryl lipid A) were disappointing with overall 20% efficacy and no efficacy against HSV-2 (Belshe *et al.*, 2012). Novel studies are now emerging that combine the gD subunit vaccine with HSV-2 immune evasion molecules glycoprotein gC2 and gE2. Addition of these two molecules to the vaccine improved efficacy in mice resulting in 97% of protection of dorsal root ganglia from HSV-2 infection at days 2-7 post challenge compared to 82% in the gD2 group only (Awasthi *et al.*, 2014).

Live vaccine: Several candidate vaccines were developed that consisted of mutant nonvirulent HSV created by deletion of one or more genes. One of them, an HSV-2 mutant deficient in glycoprotein gH, known as DISC (disabled infectious single cycle) had promising results in animal models but was clinical endpoints were not met in phase II trials that was assessing its efficacy as a therapeutic vaccine and further development was halted (McLean *et al.*, 1994).

Another mutant virus with deleted viral protein involved in Th2 polarizing response, ICP10PK rendered the virus growth and latency compromised. The mutant was effective as preventative vaccine in animal models but human clinical trials have shown only limited effects as therapeutic vaccine (Casanova *et al.*, 2002).

DNA vaccine: There have been several attempts to develop DNA-based vaccine against HSV. Currently, the Australian company Asmedus (formerly Coridon) is conducting clinical trials with its candidate DNA vaccine, developed by Dr. Ian Frazer that was shown to be 100% protective against HSV-2 in animal models (Dutton *et al.*, 2013). Phase 1 Clinical trial conducted on 20 volunteers showed that the vaccine was safe and that 3 doses were needed for optimal results. A Phase 2 clinical trial that includes a 6 months booster dose is currently ongoing and the first results are expected at the end of 2015.

CONFLICT OF INTEREST

The authors confirm that this chapter contents have no conflict of interest.

ACKNOWLEDGEMENTS

Declared None.

REFERENCES

ALLEN, S. J., HAMRAH, P., GATE, D., MOTT, K. R., MANTOPOULOS, D., ZHENG, L., TOWN, T., JONES, C., VON ANDRIAN, U. H., FREEMAN, G. J., SHARPE, A. H., BENMOHAMED, L., AHMED, R., WECHSLER, S. L. & GHIASI, H. 2011. The role of LAT in increased CD8+ T cell exhaustion in trigeminal ganglia of mice latently infected with herpes simplex virus 1. *J Virol,* 85, 4184-97.

AWASTHI, S., HUANG, J., SHAW, C. & FRIEDMAN, H. M. 2014. Blocking herpes simplex virus 2 glycoprotein E immune evasion as an approach to enhance efficacy of a trivalent subunit antigen vaccine for genital herpes. *J Virol,* 88, 8421-32.

BARINGER, J. R. 2008. Herpes simplex infections of the nervous system. *Neurol Clin,* 26, 657-74, viii.

BELSHE, R. B., LEONE, P. A., BERNSTEIN, D. I., WALD, A., LEVIN, M. J., STAPLETON, J. T., GORFINKEL, I., MORROW, R. L., EWELL, M. G., STOKES-RINER, A., DUBIN, G., HEINEMAN, T. C., SCHULTE, J. M. & DEAL, C. D. 2012. Efficacy results of a trial of a herpes simplex vaccine. *N Engl J Med,* 366, 34-43.

BERGELSON, J. M., SHAH, S S AND ZAOUTIS, T E. 2008. *Sexually Transmitted Disease. Pediatric Infectious Diseases: The Requisites in Pediatrics,* Elsevier.

CAIRNS, T. M., HUANG, Z. Y., GALLAGHER, J. R., LIN, Y., LOU, H., WHITBECK, J. C., WALD, A., COHEN, G. H. & EISENBERG, R. J. 2015. Patient-specific neutralizing antibody responses to herpes simplex virus are attributed to epitopes on either gD, gB, or both and can be type-specific. *J Virol.*

CAIRNS, T. M., LANDSBURG, D. J., WHITBECK, J. C., EISENBERG, R. J. & COHEN, G. H. 2005. Contribution of cysteine residues to the structure and function of herpes simplex virus gH/gL. *Virology,* 332, 550-62.

CAIRNS, T. M., MILNE, R. S., PONCE-DE-LEON, M., TOBIN, D. K., COHEN, G. H. & EISENBERG, R. J. 2003. Structure-function analysis of herpes simplex virus type 1 gD and gH-gL: clues from gDgH chimeras. *J Virol,* 77, 6731-42.

CARFI, A., WILLIS, S. H., WHITBECK, J. C., KRUMMENACHER, C., COHEN, G. H., EISENBERG, R. J. & WILEY, D. C. 2001. Herpes simplex virus glycoprotein D bound to the human receptor HveA. *Mol Cell,* 8, 169-79.

CASANOVA, G., CANCELA, R., ALONZO, L., BENUTO, R., MAGANA MDEL, C., HURLEY, D. R., FISHBEIN, E., LARA, C., GONZALEZ, T., PONCE, R., BURNETT, J. W. & CALTON, G. J. 2002. A double-blind study of the efficacy and safety of the ICP10deltaPK vaccine against recurrent genital HSV-2 infections. *Cutis,* 70, 235-9.

COREY, L., LANGENBERG, A. G., ASHLEY, R., SEKULOVICH, R. E., IZU, A. E., DOUGLAS, J. M., JR., HANDSFIELD, H. H., WARREN, T., MARR, L., TYRING, S., DICARLO, R., ADIMORA, A. A., LEONE, P., DEKKER, C. L., BURKE, R. L., LEONG, W. P. & STRAUS, S. E. 1999. Recombinant glycoprotein vaccine for the prevention of genital HSV-2 infection: two randomized controlled trials. Chiron HSV Vaccine Study Group. *JAMA,* 282, 331-40.

DOLAN, A., JAMIESON, F. E., CUNNINGHAM, C., BARNETT, B. C. & MCGEOCH, D. J. 1998. The genome sequence of herpes simplex virus type 2. *J Virol,* 72, 2010-21.

DUBIN, G. & JIANG, H. 1995. Expression of herpes simplex virus type 1 glycoprotein L (gL) in transfected mammalian cells: evidence that gL is not independently anchored to cell membranes. *J Virol,* 69, 4564-8.

DUTTON, J. L., LI, B., WOO, W. P., MARSHAK, J. O., XU, Y., HUANG, M. L., DONG, L., FRAZER, I. H. & KOELLE, D. M. 2013. A novel DNA vaccine technology conveying protection against a lethal herpes simplex viral challenge in mice. *PLoS One,* 8, e76407.

FATAHZADEH, M. & SCHWARTZ, R. A. 2007. Human herpes simplex virus infections: epidemiology, pathogenesis, symptomatology, diagnosis, and management. *J Am Acad Dermatol,* 57, 737-63; quiz 764-6.

FRIEDMAN, H. M. 2003. Immune evasion by herpes simplex virus type 1, strategies for virus survival. *Trans Am Clin Climatol Assoc,* 114, 103-12.

FRIEDMAN, H. M., WANG, L., FISHMAN, N. O., LAMBRIS, J. D., EISENBERG, R. J., COHEN, G. H. & LUBINSKI, J. 1996. Immune evasion properties of herpes simplex virus type 1 glycoprotein gC. *J Virol,* 70, 4253-60.

GRUNEWALD, K., DESAI, P., WINKLER, D. C., HEYMANN, J. B., BELNAP, D. M., BAUMEISTER, W. & STEVEN, A. C. 2003. Three-dimensional structure of herpes simplex virus from cryo-electron tomography. *Science,* 302, 1396-8.

HELDWEIN, E. E., LOU, H., BENDER, F. C., COHEN, G. H., EISENBERG, R. J. & HARRISON, S. C. 2006. Crystal structure of glycoprotein B from herpes simplex virus 1. *Science,* 313, 217-20.

HIDAKA, Y., SAKAI, Y., TOH, Y. & MORI, R. 1991. Glycoprotein C of herpes simplex virus type 1 is essential for the virus to evade antibody-independent complement-mediated virus inactivation and lysis of virus-infected cells. *J Gen Virol,* 72 (Pt 4), 915-21.

HOMA, F. L. & BROWN, J. C. 1997. Capsid assembly and DNA packaging in herpes simplex virus. *Rev Med Virol,* 7, 107-122.

HUTCHINSON, L., BROWNE, H., WARGENT, V., DAVIS-POYNTER, N., PRIMORAC, S., GOLDSMITH, K., MINSON, A. C. & JOHNSON, D. C. 1992. A novel herpes simplex virus glycoprotein, gL, forms a complex with glycoprotein H (gH) and affects normal folding and surface expression of gH. *J Virol,* 66, 2240-50.

JENNINGS, S. R., BONNEAU, R. H., SMITH, P. M., WOLCOTT, R. M. & CHERVENAK, R. 1991. CD4-positive T lymphocytes are required for the generation of the primary but not the secondary CD8-positive cytolytic T lymphocyte response to herpes simplex virus in C57BL/6 mice. *Cell Immunol,* 133, 234-52.

KEIR, M. E., BUTTE, M. J., FREEMAN, G. J. & SHARPE, A. H. 2008. PD-1 and its ligands in tolerance and immunity. *Annu Rev Immunol,* 26, 677-704.

KELLY, B. J., FRAEFEL, C., CUNNINGHAM, A. L. & DIEFENBACH, R. J. 2009. Functional roles of the tegument proteins of herpes simplex virus type 1. *Virus Res,* 145, 173-86.

KHANNA, K. M., LEPISTO, A. J., DECMAN, V. & HENDRICKS, R. L. 2004. Immune control of herpes simplex virus during latency. *Curr Opin Immunol,* 16, 463-9.

KNICKELBEIN, J. E., KHANNA, K. M., YEE, M. B., BATY, C. J., KINCHINGTON, P. R. & HENDRICKS, R. L. 2008. Noncytotoxic lytic granule-mediated CD8+ T cell inhibition of HSV-1 reactivation from neuronal latency. *Science,* 322, 268-71.

LANG, A. & NIKOLICH-ZUGICH, J. 2005. Development and migration of protective CD8+ T cells into the nervous system following ocular herpes simplex virus-1 infection. *J Immunol,* 174, 2919-25.

LEHMAN, I. R. & BOEHMER, P. E. 1999. Replication of herpes simplex virus DNA. *J Biol Chem,* 274, 28059-62.

LIU, T., KHANNA, K. M., CARRIERE, B. N. & HENDRICKS, R. L. 2001. Gamma interferon can prevent herpes simplex virus type 1 reactivation from latency in sensory neurons. *J Virol,* 75, 11178-84.

LIU, T., KHANNA, K. M., CHEN, X., FINK, D. J. & HENDRICKS, R. L. 2000. CD8(+) T cells can block herpes simplex virus type 1 (HSV-1) reactivation from latency in sensory neurons. *J Exp Med,* 191, 1459-66.

LONG, S. S., PICKERING, L K AND PROBER, C G. 2009. *Principles and Practice of Pediatric Infectious Diseases.*, Elsevier.

MANDELL, G. L., BENNETT, J E AND DOLLIN, R. 2005. *Principles and Practice of Infectious Diseases.*, Philadelphia, Elsevier.

MCLEAN, C. S., ERTURK, M., JENNINGS, R., CHALLANAIN, D. N., MINSON, A. C., DUNCAN, I., BOURSNELL, M. E. & INGLIS, S. C. 1994. Protective vaccination against primary and recurrent disease caused by herpes simplex virus (HSV) type 2 using a genetically disabled HSV-1. *J Infect Dis,* 170, 1100-9.

MCNEARNEY, T. A., ODELL, C., HOLERS, V. M., SPEAR, P. G. & ATKINSON, J. P. 1987. Herpes simplex virus glycoproteins gC-1 and gC-2 bind to the third component of complement and provide protection against complement-mediated neutralization of viral infectivity. *J Exp Med,* 166, 1525-35.

MOTT, K. R., BRESEE, C. J., ALLEN, S. J., BENMOHAMED, L., WECHSLER, S. L. & GHIASI, H. 2009. Level of herpes simplex virus type 1 latency correlates with severity of corneal scarring and exhaustion of CD8+ T cells in trigeminal ganglia of latently infected mice. *J Virol,* 83, 2246-54.

MUELLER, S. N. & AHMED, R. 2009. High antigen levels are the cause of T cell exhaustion during chronic viral infection. *Proc Natl Acad Sci U S A,* 106, 8623-8.

MURRAY, P. R., ROSENTHAL, K S AND PFALLER, M A. 2008. *Medical Microbiology,* Elsevier.

NEWCOMB, W. W., THOMSEN, D. R., HOMA, F. L. & BROWN, J. C. 2003. Assembly of the herpes simplex virus capsid: identification of soluble scaffold-portal complexes and their role in formation of portal-containing capsids. *J Virol,* 77, 9862-71.

PENG, T., PONCE DE LEON, M., NOVOTNY, M. J., JIANG, H., LAMBRIS, J. D., DUBIN, G., SPEAR, P. G., COHEN, G. H. & EISENBERG, R. J. 1998. Structural and antigenic analysis of a truncated form of the herpes simplex virus glycoprotein gH-gL complex. *J Virol,* 72, 6092-103.

REMINGTON, J. S. E. A. 2006. *Infectious Diseases of the Fetus and Newborn Infant.,* Elsevier.

RESKE, A., POLLARA, G., KRUMMENACHER, C., CHAIN, B. M. & KATZ, D. R. 2007. Understanding HSV-1 entry glycoproteins. *Rev Med Virol,* 17, 205-15.

RUOCCO, E., DONNARUMMA, G., BARONI, A. & TUFANO, M. A. 2007. Bacterial and viral skin diseases. *Dermatol Clin,* 25, 663-76, xi.

SALIO, M., CELLA, M., SUTER, M. & LANZAVECCHIA, A. 1999. Inhibition of dendritic cell maturation by herpes simplex virus. *Eur J Immunol,* 29, 3245-53.

SCHOLTES, L. & BAINES, J. D. 2009. Effects of major capsid proteins, capsid assembly, and DNA cleavage/packaging on the pUL17/pUL25 complex of herpes simplex virus 1. *J Virol,* 83, 12725-37.

SMITH, J. S. & ROBINSON, N. J. 2002. Age-specific prevalence of infection with herpes simplex virus types 2 and 1: a global review. *J Infect Dis,* 186 Suppl 1, S3-28.

SPEAR, P. G., EISENBERG, R. J. & COHEN, G. H. 2000. Three classes of cell surface receptors for alphaherpesvirus entry. *Virology,* 275, 1-8.

TAKAI, Y. & NAKANISHI, H. 2003. Nectin and afadin: novel organizers of intercellular junctions. *J Cell Sci,* 116, 17-27.

VITTONE, V., DIEFENBACH, E., TRIFFETT, D., DOUGLAS, M. W., CUNNINGHAM, A. L. & DIEFENBACH, R. J. 2005. Determination of interactions between tegument proteins of herpes simplex virus type 1. *J Virol,* 79, 9566-71.

WARNER, M. S., GERAGHTY, R. J., MARTINEZ, W. M., MONTGOMERY, R. I., WHITBECK, J. C., XU, R., EISENBERG, R. J., COHEN, G. H. & SPEAR, P. G. 1998. A cell surface protein with herpesvirus entry activity (HveB) confers susceptibility to infection by mutants of herpes simplex virus type 1, herpes simplex virus type 2, and pseudorabies virus. *Virology,* 246, 179-89.

WATSON, R. J. & VANDE WOUDE, G. F. 1982. DNA sequence of an immediate-early gene (IEmRNA-5) of herpes simplex virus type I. *Nucleic Acids Res,* 10, 979-91.

WHITBECK, J. C., PENG, C., LOU, H., XU, R., WILLIS, S. H., PONCE DE LEON, M., PENG, T., NICOLA, A. V., MONTGOMERY, R. I., WARNER, M. S., SOULIKA, A. M., SPRUCE, L. A., MOORE, W. T., LAMBRIS, J. D., SPEAR, P. G., COHEN, G. H. & EISENBERG, R. J. 1997. Glycoprotein D of herpes simplex virus (HSV) binds directly to HVEM, a member of the tumor necrosis factor receptor superfamily and a mediator of HSV entry. *J Virol,* 71, 6083-93.

WHITLEY, R. J., KIMBERLIN, D. W. & ROIZMAN, B. 1998. Herpes simplex viruses. *Clin Infect Dis,* 26, 541-53; quiz 554-5.

ZHOU, Z. H., CHEN, D. H., JAKANA, J., RIXON, F. J. & CHIU, W. 1999. Visualization of tegument-capsid interactions and DNA in intact herpes simplex virus type 1 virions. *J Virol,* 73, 3210-8.

Immune Responses to CMV and Vaccine Development

Masha Fridkis-Hareli[*]

ATR LLC, Sudbury, MA 01776, USA

Abstract: Cytomegalovirus (CMV) is found throughout the world in all geographic and socioeconomic groups, but, in general, it is more widespread in developing countries and in areas of lower socioeconomic conditions. CMV still remains a major human pathogen causing significant morbidity and mortality in immunosuppressed or immunocompromized individuals. Between 50% and 80% of adults in the United States are infected with CMV by 40 years of age. CMV is the most common congenitally transmitted virus, resulting in approximately 1 in 150 children born with congenital CMV *infection, and in* approximately 1 in 750 children developing *permanent* disabilities *due to CMV*. Thus, development of vaccines against CMV infections has been a major biomedical research priority. In this chapter, immunobiology of host-CMV interactions is discussed as related to the host immune responses against viral infection, providing insights into the complex interplay between host and virus that facilitates viral persistence. In addition, an update on CMV vaccines that are currently in preclinical and clinical development, is provided, outlining important questions about the nature of protective immune responses that will be required for potential CMV immunization strategies.

Keywords: Adoptive transfer, anti-viral drugs, anti-viral immune responses, cell-mediated immunity, CMV, congenital transmission, host-virus interactions, immunocompromised, immunodominant, immunosuppression, immunization, primary infection, prophylaxis, seropositive, symptomatic disease, T cells, vaccines, viral clearance, viremia, virulence

INTRODUCTION

Cytomegalovirus (CMV) is a member of Betaherpesvirinae in the subfamily Herpesviridae that also includes human herpesvirus (HHV)–6 and HHV-7 that share common clinical characteristics with CMV (Huang, *et al.*, 1980). Most people are infected with CMV at some stage in their life. While in developing countries most infections are acquired during childhood, in developed countries

*Corresponding author Masha Fridkis-Hareli: ATR LLC, Sudbury, MA 01776, USA;
E-mail: mhareli@bio-atr.com

Liljana Stevceva (Ed)

up to 50% of young adults are CMV seronegative. In immunocompetent individuals, CMV is usually an asymptomatic infection but it sometimes manifests clinically as mononucleosis. Clinically significant CMV disease frequently develops in patients immunocompromised by advanced HIV infection, post solid-organ transplantation immunosuppressive therapy, and post bone-marrow transplantation immunosuppressive therapy (Freeman, 2009). Additionally, congenital transmission of CMV from a mother acutely infected during pregnancy may cause neurological abnormalities and deafness in the newborn. Symptomatic disease in immunocompromised individuals may present clinically as fever, pneumonia, hepatitis, encephalitis, myelitis, colitis, uveitis, retinitis, and neuropathy. CMV, similarly to other herpesviruses, establishes a latent infection in the host that may reactivate during periods of immunosuppression.

Multiple genetically distinct strains of CMV with differences in genotypes may be associated with variations in virulence (Huang, *et al.*, 1980). Infection with more than one strain could be a possible explanation for the cases of congenital CMV in children of seropositive mothers and has been observed in organ transplant patients. CMV has the largest genome of the herpes viruses, ranging from 230-240 kilobase pairs and is arranged as a double-stranded linear DNA with 162 hexagonal protein capsomeres surrounded by a lipid membrane. The outer envelope of the virus, which is derived from the host cell nuclear membrane, contains multiple virally encoded glycoproteins. Glycoprotein B (gB) and glycoprotein H (gH) appear to be the major determinants of protective humoral immunity. Analysis of the coding sequences of gB and gH showed clustered variations. Clinical CMV isolates were found to adopt one of four gB and two gH sequence configurations at certain loci (Fries, *et al.*, 1994). This genetic variation in gB was found to affect the viral pathogenicity and clinical outcome in immunocompromised patients (Fries, *et al.*, 1994).

In primary infection, CMV immunoglobulin (Ig) M antibodies may be found as early as 4-7 weeks and may persist as long as 16-20 weeks after initial infection. The majority of neutralizing antibody is directed against an envelope glycoprotein gB. Studies have shown that more than 50% of neutralizing activity in convalescent serum is attributable to glycoprotein gB. However, virion tegument proteins such as pp150, pp28, and pp65 evoke strong and durable antibody responses. Cell-mediated immunity is considered the most important factor in controlling CMV infection. Patients deficient in cell-mediated immunity are at greatest risk for CMV disease. CMV-specific CD4[+] and CD8[+] lymphocytes play an important role in immune protection after primary infection or reactivation of latent disease (van Leeuwen, *et al.*, 2006). Studies in bone marrow transplant patients have revealed that patients who do not develop CMV-specific CD4$^+$ or CD8$^+$ cells are at higher risk for CMV pneumonitis. Additionally, no cases of CMV pneumonia have been reported in immunosupressed allogeneic marrow transplant patients receiving infusions of CMV-specific CD8$^+$ cells (Barry *et al.*, 2000).

Primary CMV infection is usually asymptomatic or induces mild flulike symptoms. The immunocompromised host, however, carries the greatest risk for CMV disease. CMV excretion in the saliva and urine is common in patients who are immunocompromised and is generally of little consequence while viremia identifies those at greatest risk for clinically manifested CMV disease (Humar, *et al.*, 2010).

Prophylactic or presymptomatic antiviral therapy against CMV disease in transplant recipients typically relies on the detection of CMV in the blood by shell vial cultures, CMV antigenemia, CMV pp65 or pp67 antigen assays, and PCR amplification, which is the most sensitive assay (Miller, *et al.*, 2010). There are two types of PCR tests: the viremia or qualitative assays that determine if a person is CMV-positive or -negative and the quantitative assays that measure viral load. Hoffman-La Roche and BioSource International have both developed these assays, which are available through certain labs (Biosource is offering both tests, while Roche is just offering the qualitative). Traditionally, CMV antibody tests were performed using complement fixation and showed peak viral titers 4-7 weeks after infection. Multiple tests for CMV antibody are now available. Some tests are sensitive enough to detect anti-CMV IgM antibody early in the course of the illness and during CMV reactivation. Reactivation of the virus is not uncommon, sometimes occurring with viremia and a positive IgM in the presence of IgG antibody. This is usually observed during intercurrent infections or at times of patient stress. The clinical significance, time course, and natural history of reactivation in immunocompetent patients are not known for either of the viruses.

The CMV immune status of the mother is important in determining the risk of placental infection and subsequent symptomatic disease in the child or fetus. Symptomatic CMV congenital disease is less likely to occur in women with pre-existing immune responses to CMV than in CMV-naïve individuals (Boppana, *et al.*, 2010). One in ten cases of acute CMV during pregnancy is estimated to result in congenital CMV disease carrying a risk of significant symptomatic disease and developmental defects in newborns (Boppana, *et al.*, 2010). The most common clinical findings include hyperbilirubinemia (81%), increased levels of hepatocellular enzymes (83%), thrombocytopenia (77%), and increased CSF protein levels (77%). Studies have shown that asymptomatic children with neurological findings are more likely to have CMV IgM antibody. Many cases of hearing loss in children may be caused by CMV infection.

Adults manifesting CMV infection as a mononucleosis syndrome may occasionally develop pneumonia occurring at a rate of approximately 0-6%. It rapidly resolves with the disappearance of the primary infection. Clinically significant and life-threatening CMV pneumonia may develop in immunocompromised patients (Scalzo, *et al.*, 2007; Piñana, *et al.*, 2010). Those most at risk are bone-marrow transplant patients and recipients of lung transplants. In patients who have received marrow transplants, CMV disease is most likely 30-60 days after transplant. The differential diagnoses in patients who are immunocompromised include Pneumocystis pneumonia, respiratory viruses, pulmonary hemorrhage, drug toxicity, recurrent lymphoma, and other infections. Notably, CMV is frequently detected in the lungs of patients with HIV/AIDS but does not frequently cause clinically significant disease.

ROLE OF HOST AND VIRAL FACTORS INFLUENCING THE OUTCOME OF CMV INFECTION

Although significant advances have been achieved in studying the host response to human CMV infection, the strict species specificity of CMVs suggests that most aspects of antiviral immunity are best assessed in animal models. The mouse model of murine CMV infection has been an important tool for analysis of *in vivo* features of host–virus interactions and responses to antiviral drugs that are difficult to assess in humans. Important studies of the contribution of host resistance genes to infection outcome, interplays between innate and adaptive host immune responses, the contribution of virus immune evasion genes and genetic variation in these genes to the establishment of persistence and *in vivo* studies of resistance to antiviral drugs have benefited from the well-developed murine CMV model (Chalmer, Mackenzie and Stanley, 1977).

The outcome of host–virus interactions is determined by a range of factors including inherent host resistance, environmental factors, such as nutritional status, and viral factors such as variation in genes governing levels of infectivity, tropism and immune evasion. Different individuals vary in their relative susceptibility to infection because some host genes that confer resistance are polymorphic between individuals within the same population. The outbred nature of the human population has made it difficult to assess the contribution of host resistance genes in determining disease outcome of human CMV infection. However, murine CMV infection of inbred mouse strains provided many important insights into host genes that regulate CMV infection (Price, *et al.*, 1987; 1990). To investigate the contribution of the *H2* complex to MCMV resistance, Price *et al.* isolated macrophages from mice differing in *H2* genotype and assessed their susceptibility to MCMV infection *in vitro*. These studies showed that cells from $H2^k$ haplotype mice were considerably less sensitive to infection, and mapping studies using intra-*H2* haplotype mice indicated that MHC class I genes contributed to this effect (Price, *et al.*, 1987; 1990; Wykes, *et al.*, 1993). Further evidence for a role of MHC class I molecules was demonstrated by the finding that transfection of cells that were largely resistant to CMV infection with specific MHC class I molecules increased their susceptibility to CMV infection (Wykes, *et al.*, 1993). Altogether, these data indicated that MHC class I molecules could function as a receptor or coreceptor for CMV entry.

The importance of NK cells in controlling murine CMV was demonstrated about two decades ago in studies describing exacerbation of the CMV infection when adult B6 mice were depleted of NK cells or amelioration when NK cells were adoptively transferred into suckling mice before CMV infection. Analyses of NK cell activity following CMV infection in inbred strains also revealed a general correlation between the level of NK cell cytotoxicity and resistance status to CMV (Bancroft, Shellam and Chalmer, 1981; Bukowski, Woda and Welsh, 1984).

Beutler, B *et al.* introduced the term 'resistome' underlying resistance to infection that is largely inherited rather than acquired, and is encoded by a definable set of host genes into which spontaneous or induced germline mutations have been introduced. Mutations induced by random germline mutagenesis have become numerous, enabling to define the size of the resistome and the understanding of how they interact. N-ethyl-N-nitrosourea mutagenesis effort, which recently showed that components of Toll-like receptor signaling are essential constituents of the arsenal against MCMV infections, validated the forward genetic approach as a powerful tool to define the resistome (Beutler, *et al.*, 2005; Crozat, *et al.*, 2006).

CMV-SPECIFIC T CELLS: PHENOTYPIC AND FUNCTIONAL CHARACTERISTICS

Studies of individuals after primary virus infections have revealed how divergent virus-specific CD8$^+$ T cells may develop from the initially expanded virus-specific T-cell effector pool. Many persistent virus-specific T cells, recognizing CMV, lack IL-7 receptor α (IL-7Rα) and depend on viral antigens to persist. CMV is unique in that it generates a vast pool of resting virus-specific T cells with constitutive cytolytic effector function (Fries, *et al.*, 1994).

Current view on the regulation of expansion and maintenance of virus-specific T cells in response to infecting pathogens obtained from studies in animals states that antigen-specific CD8$^+$ T cells proliferate to form a large pool of effector cells that are capable of fighting the infection, but once the pathogen has been cleared, most effector cells die in the so-called contraction phase, and around 10% remain that form the long-lived memory population (Welsh, Selin and Szomolanyi-Tsuda, 2004). In humans, the situation is different because the exact time point of infection is not known and symptoms may not be recognized or only develop weeks after the infection. In addition, persistence of latent viruses in humans requires presence of active T cells, *e.g.*, the first wave of CMV infection occurs during early childhood, which means that the immune system must have the capacity to maintain CMV latency for more than 80 years. Moreover, many of the relevant viruses are not eliminated but are persistent and have to be continuously controlled by the immune system. Thus, in humans, active anti-CMV specific T cells represent a larger size circulating pool compared to animals. Tetramer technology (Altman, *et al.*, 1996) has been widely used to address issues related to the induction and maintenance of anti-CMV CD8$^+$ T-cell responses, diversity of human virus-specific T-cell responses, and properties of tissue-residing T cells (van Leeuwen, *et al.*, 2006).

During the early phase of the antiviral response to acute infection, the vast majority of circulating virus-specific T cells are activated and in cell cycle, as reflected by the cell-surface expression of the activation markers CD38 and HLA-DR and the intracellular presence of Ki67, a nuclear antigen found in dividing cells. Early CMV-specific cytotoxic T cells express perforin and granzyme B, and therefore mirror the acutely formed murine effector T cells (Gamadia, *et al.*, 2003). Apart from the expression of the above-mentioned activation markers, these first CMV-specific CD8$^+$ T cells are noticeably different from unprimed cells, expressing CD45RO isoform of CD45 as opposed to CD45RA on naïve CD8$^+$ T cells (van Leeuwen, *et al.*, 2006). The CD45RA^{neg}CD45R0^{pos} phenotype of the activated CMV-reactive cells is in agreement with *in vitro* studies, showing that T-cell activation induces a shift from CD45RA to CD45R0 expression (Smith, *et al.*, 1986). Notably, CD62L (Chen, 2001) and CCR7 (Gamadia, *et al.*, 2003) are not expressed by CMV-specific CD8+ T cells, suggesting that after priming in the secondary lymphoid tissue by activated antigen-presenting cells (APCs), virus-reactive cells migrate from the lymph node compartment toward the infected tissue to contain virus replication. It has been shown recently that a minority of virus-specific CD8+ effector T cells express IL-7Rα after acute infection with a cleared pathogen (van Leeuwen, *et al.*, 2006). Still, this small fraction seeds the memory pool, as these cells can respond to the homeostatic cytokine IL-7, when the antigen is eliminated. The early CMV-specific CD8$^+$ T-cell population is completely devoid of IL-7Rα-expressing cells, which parallels the lack of this receptor on T cells recognizing persistent viruses in mice (Wherry, *et al.*, 2004). Early CMV-specific T cells express the co-stimulatory receptors CD28 and CD27.

After efficient control of viral replication is established and the CMV-DNA levels in whole blood become undetectable, the phenotype of the specific T-cell pool continues to change in the following months. Gradually, in the first months after infection CMV-specific T cells lose the expression of CD28. With somewhat slower kinetics, CD27 disappears, and approximately 1 year after infection, CMV-specific T cells are either CD27dull or CD27^{neg} (Gamadia, *et al.*, 2003). The CD27 protein is a member of the TNF receptor superfamily and it binds to the ligand CD70. CD27 plays and important role in inducing and maintaining T cell immunity and through its ligand CD70 in B cell activation and immunoglobulin synthesis.

IL-7Rα-expressing cells, which are absent early in infection, emerge in the circulation when the CMV-DNA load becomes undetectable. Lastly, a considerable number of cells appear to switch from CD45R0 back to CD45RA expression. It should be noted that CMV infection induces a dramatic change in the appearance of the total CD8$^+$ population. Before infection, the majority of T cells express CD28, CD27, and IL-7Rα. After infection, this population is significantly reduced, and a CD28^{neg}CD27^{neg}IL-7Rα^{neg} T cell subset appears. This subset is also present in healthy CMV-carrying individuals and indicates that CMV infection has major impact on the immune system (Kuijpers, *et al.*, 2003; Sylwester, *et al.*, 2005). It is unlikely that CMV-specific cells are specifically kept within the circulating pool, because considerable numbers can be detected in human spleen (Langeveld, Gamadia and ten Berge, 2006). Moreover, CMV infection was found to increase total CD8$^+$ T-cell numbers, suggesting that CMV infection does not impose a competition within the CD8$^+$ T-cell compartment but rather provokes the generation of a particular fraction of class I-restricted cells devoted to maintaining CMV latency.

The development of CD27^{neg} CMV-specific T cells infers that CMV infection induces ample CD70 expression in the infected patient, and indeed during active CMV replication, CD70-expressing T cells can be found in the circulation (Gamadia, *et al.*, 2004). The amount of CD70 that is induced on *in vitro* activated T cells is related to the strength of the TCR signal, but no data are available on numbers of CD70-expressing cells *in vivo* during CMV replication. However, the percentage of CD27^{neg} CMV-specific cells negatively correlated with peak viral loads, suggesting that strong viral replication induces high levels of CD70 and consequently induces the emergence of a large fraction of CD27^{neg} CMV-specific cells. Interestingly, the amount of antigen also has a direct effect on the size of the virus-specific T-cell compartment, and therefore, a significant positive correlation has been found between the percentage of CD27^{neg} cells and the magnitude of the CMV-directed CD8$^+$ T cells response during latency (Papagno, *et al.*, 2002; Gamadia, *et al.*, 2004).

The mechanism that maintains most CMV-reactive T cells in an IL-7Rα^{low} phenotype is unknown, but the observation that an initial high viral load correlates with a high percentage of IL-7Rα^{low} cells during latency may suggest that this phenotype already exists early in the antiviral response. Whether the minor fraction of IL-7Rα^{pos} CMV-specific CD8$^+$ T cells are derived from the early IL 7Rα^{low} pool or rather are novel cells that emerge once the virus has entered the latent stage will have to be further investigated. Altogether, these findings lead to the hypothetical model described in van Leeuwen *et al.* (2006). In the situation of infection with a persistent virus, the IL-7Rα^{neg} T cells can survive, because they are regularly triggered by antigen and therefore do not depend on IL-7 for survival. This model would explain why memory T cells specific for viruses that have been cleared all express the IL-7Rα. Simultaneously, it would clarify the finding that a higher viral load results in lower percentages of IL-7Rα^{pos} CMV-specific cells (van Leeuwen, *et al.*, 2006).

The CD45RA^{pos}CCR7^{neg}CD28^{neg}CD27^{neg} phenotype is frequently found for CMV-specific cells (Kern, *et al.*, 1999; Sandberg, Fast and Nixon, 2001). Recent findings suggested that the expression of IL-7Rα might be an additional marker to subdivide sets of virus-specific T cells (Boutboul, *et al.*, 2005). The fraction of virus-specific IL-7Rα^{low} cells was reported to increase during CMV reactivation and then decreased again (van Leeuwen, *et al.*, 2006). The progression through the T cell subsets based on the expression of CD27 and CD28 might be related to the chronicity of antigen exposure, as CD8$^+$ T cells specific for cleared viruses have an early phenotype, contrasting with cells specific for the frequently reactivating CMV that are in the late differentiation subset.

The CD8+CD45RA^{pos}CCR7^{neg}CD28^{neg}CD27^{neg} T-cell population (named effector-type or late memory) is the only subset in the circulation that can execute immediate virus-specific effector functions in donors without any clinical signs of acute viral disease. The size of this subset is strongly correlated with CMV seropositivity but not with previous exposure to other persisting herpesviruses such as EBV or varicella zoster virus, which concurs with the observation that CMV-specific CD8^{+} T cells frequently reach a CD45RA^{pos}CD27^{neg} phenotype (van Leeuwen, *et al.*, 2006). The size of this population increases with age and during immunosuppression, and in healthy CMV-carrying adults, the frequency of this population in the circulating CD8^{+} T-cell pool may reach over 30% (van Leeuwen, *et al.*, 2006). This observation, together with recent findings of very high frequencies of CMV-specific T cells in CMV-seropositive donors (Boutboul, *et al.*, 2005), shows that CMV infection has a major impact on the CD8^{+} T-cell compartment.

In vivo observations show that during CMV reactivation, the CMV-reactive CD45RA^{pos}CD27^{neg} T-cell population becomes activated, as evidenced by the high expression of both CD38 and HLA-DR (Gamadia, *et al.*, 2004; Sylwester, *et al.*, 2005). In correspondence with the *in vitro* data, the cells switch from CD45RA to CD45R0 but retain their CD27^{neg} phenotype. Importantly, the CMV-specific pool expands during and after reactivation, showing that this population can respond with renewed clonal expansion to increased viral load. Regarding their role in achieving and maintaining viral latency, Cobbold, *et al.* (2005) recently showed that tetramer-selected CMV-specific T cells were able to reduce viral loads after adoptive transfer into stem cell transplant recipients with CMV viremia. These data argue that CD45RA^{pos}CD27^{neg} T cells are effective in mediating strong antiviral responses *in vivo*. In summary, CD8^{+}CD45RA^{pos}CD27^{neg} T cells are actively involved in the suppression of viral replication in persistent viral infections.

CD4^{+} T cells also play a crucial role in the control of HCMV infection. Kern *et al.* (2002), examining the response of 40 donors to peptides derived specifically from the pp65 antigen, showed that 63% of normal healthy donors have a CD4 T cell response and 83% have a CD8 T cell response, indicating that this protein is an important target for both CD4 and CD8 T cells and that the response to CMV is high in the majority of individuals. The importance of CD4 T cells is highlighted further in a study examining the kinetics and characteristics of CMV-specific CD4^{+} and CD8^{+} T cells in the course of primary CMV infection in patients receiving renal transplants (Gamadia, *et al.*, 2003). The authors showed that in asymptomatic individuals the CMV-specific CD4$^+$ T cell response preceded the CMV-specific CD8$^+$ T cell response; however, in symptomatic patients the CD4$^+$ T cell response was delayed and detected only after anti-viral treatment. These findings imply that the presence of functional and specific CD8$^+$ T cell and antibody responses are not sufficient to control viral replication and that the formation of specific effector CD4$^+$ T cells is essential for clearance of infection (Gamadia, *et al.*, 2003). Prolonged viral shedding in urine and saliva for at least 12–29 months after acquisition of CMV occurs in immunocompetent young children who acquire CMV (6 months in adults). This correlates with decreased CMV-specific T helper 1 (Th1) response, as measured by the secretion of IFNγ and IL-2 (Tu, *et al.*, 2004). The authors suggest that CD4$^+$ T cell immunity to HCMV may be generated in an age-dependent manner (Tu, *et al.*, 2004).

ADVANCES IN CMV VACCINE DEVELOPMENT

Development of a vaccine against CMV infection, and in particular, against congenital CMV infection and disease, has been a major biomedical research priority (Schleiss, 2005, 2008). In addition, solid organ transplant and haematopoietic stem cell transplant patients could also benefit from vaccination against CMV. Currently, the various CMV vaccines evaluated in preclinical and clinical trials include recombinant protein subunit vaccines, poxvirus and alphavirus-vectored subunit vaccines, DNA vaccines, live and attenuated vaccines, dense body vaccines and passive vaccine strategies, based on adoptive transfer of CMV-specific T-cells and neutralizing IgG. However, there is uncertainty as to the optimal patient populations to target for vaccination. The most appropriate approach to the implementation of CMV vaccines into clinical practice may depend on the patient population being protected.

Women of childbearing age and individuals undergoing immunosuppressive treatment prior to transplantation are considered to be potential target populations for the development and utilization of CMV vaccines. Arguably the most compelling rationale for developing a vaccine against CMV is the prevention of disease resulting from congenital CMV infection. CMV is the most common congenitally transmitted viral pathogen encountered in newborns in the developed world, and it is estimated that congenital CMV transmission occurs in 0.5–2% of all newborns (Strangert, *et al.*, 1976; Stanberry, *et al.*, 2004). Congenital infections are more common in CMV-seropositive women, young mothers and in infants born to women from lower socio-economic backgrounds. In the USA, it is currently estimated that approximately 40,000 newborn infants are infected annually, calling for the need to increase public awareness about congenital CMV infection and to promote vaccine development.

The economic costs to the society that is associated with congenital CMV infection are considerable. In the early 1990s, analysis of the disease burden associated with congenital CMV infection estimated a cost to the US healthcare system of approximately US $1.9 billion annually, and a cost per affected child of over US $300,000, which reflects the lifelong disability associated with symptomatic infection, since patients often require long-term residential care and extensive medical intervention (Whitley, 2004). There are few options available for ameliorating the neurodevelopmental injury associated with congenital CMV infection, with preconceptual vaccination being the most useful interventional measure for preventing the sequelae of congenital CMV. Observations of the generally protective effect of maternal immunity on CMV transmission and subsequent CMV disease in newborns support the concept of development of a CMV vaccine that is targeted primarily at young women. In addition, programs which target universal immunization against CMV in early childhood may have the potential to confer a lifetime of benefit for an individual.

Immunocompromised patients, in particular transplant patients, are at high risk of CMV disease that may be clinically manifested as a variety of conditions, including pneumonitis, colitis and CMV syndrome (Demmler, 1996). Effective prophylactic and pre-emptive therapy has made CMV a rare cause of mortality in stem cell transplant recipients. However, CMV-seropositive transplant recipients continue to have a considerable and persistent mortality disadvantage when compared with CMV-seronegative recipients with a seronegative donor. Graft survival for solid transplant recipients is also negatively affected by CMV status (Porath, *et al.*, 1990). Prevention strategies that employ vaccines capable of stimulating both humoral and cell-mediated immune responses to CMV may therefore be of value in further decreasing the incidence (and severity) of CMV disease post-transplantation. Such vaccines could be administered to either the transplant recipient or to the stem cell or bone marrow donor prior to transplantation. Potential benefits could include reduced CMV disease following transplantation, reduced use of antiviral therapy, prolonged graft survival and reduction in CMV-associated transplant complications, including graft-*versus*-host disease and fungal infections.

Several CMV vaccines have been evaluated in preclinical and clinical studies. The first live, attenuated CMV vaccine candidate tested in humans was derived from the AD169 strain of CMV, a laboratory-adapted strain which was modified by passage of an isolate (first cultured from human adenoidal tissue) 54 times in human fibroblasts (Boeckh, Fries and Nichols, 2004). This vaccine was found to be safe and generally well tolerated when administered to CMV-seronegative adults, with the exception of common injection site reactions and mild systemic symptoms. The majority of seronegative adults inoculated with AD169 vaccine developed CMV-specific antibodies. Participants with pre-existing immunity to CMV exhibited no augmentation of antibody response to vaccination. Subsequently, the CMV Towne strain was developed as a potential live, attenuated vaccine candidate. The initial human trial with the Towne vaccine yielded similar results to those obtained in the AD169 trials. All 10 seronegative adults inoculated intramuscularly with Towne seroconverted within 4 weeks of vaccination, whereas none of the five vaccinated persons with pre-existing CMV immunity developed an augmented antibody response (Griffiths, 2003). In an attempt to establish a non-invasive vaccination route, 11 CMV-seronegative adults were inoculated both intranasally and orally with Towne; however, none developed CMV-specific antibodies. Subsequent human trials with the Towne vaccine confirmed its ability to elicit antibodies with similar specificities to antibodies induced by natural CMV infection (Elek and Stern, 1974). The Towne vaccine also engenders cell-mediated immune responses: healthy seronegative adults receiving this vaccine uniformly develop CMV-specific lymphoproliferative responses, persisting for at least 10 months post-vaccination; Towne vaccination also consistently elicits CMV-specific $CD8^+$ cytotoxic T-lymphocyte responses in immunocompetent individuals (Just, *et al.*, 1975; Gonczol, *et al.*, 1989). Moreover, the Towne vaccine is the only CMV vaccine candidate to undergo efficacy testing in prospective kidney transplant recipients. Evaluations in this population revealed that most renal transplant candidates developed humoral and cellular immune responses to Towne vaccination. In addition, in the highest risk population, the incidence of severe CMV disease was reduced by 72–100%, a degree of protection comparable with that conferred by natural infection. These studies suggested that the Towne vaccine is safe and well tolerated in CMV-seronegative and CMV-seropositive people, and induces both humoral and cellular immunity to the virus. Furthermore, the safety of the vaccine was demonstrated by the absence of systemic symptoms or clinical laboratory abnormalities, no evidence for latency after immunosuppression, injection site reaction common without systemic side-effects and no depression of cell-mediated immunity or alteration of CD4/CD8 ratio. Immunogenicity was shown by eliciting humoral response including neutralizing antibody, induction of lasting lymphocyte proliferative responses and of HLA-restricted cytotoxicity (Plotkin,

2000). In attempt to augment the efficacy, 3000 pfu Towne CMV vaccine, with or without adjuvant recombinant interleukin-12 (rhIL-12), were administered to CMV-seronegative healthy volunteers and then measured CMV gB-specific IgG titers and CMV-specific $CD4^+$ and $CD8^+$ T cell proliferation and IFNγ expression after stimulation with whole viral lysate and immunodominant peptide CMV antigens (Jacobson, *et al.*, 2006). Adjuvant rhIL-12 at doses up to 2 μg were well-tolerated and associated with (a) dose-related increases in peak anti-CMV gB IgG titers (though not in sustained titers), (b) dose-related increases in the weak CMV viral lysate-specific $CD4^+$ T cell proliferation responses induced by vaccine alone after 360 days of follow-up, and (c) decreases in the very robust CMV IE-specific peak $CD4^+$ T cell and Day 360 $CD8^+$ T cell proliferation responses induced by the vaccine alone. Also, qualitative $CD8^+$ T cell IFNγ responses to stimulation with the immunodominant CMV antigen, pp65, tended to occur more frequently in vaccinees who received 0.5-2.0 μg rhIL-12 compared to lower dose or no rhIL-12. Thus, adjuvant IL-12 may be a promising strategy for improving antibody and T cell immune responses to a CMV vaccine (Starr, *et al.*, 1981; Quinnan, *et al.*, 1984; Jacobson, *et al.*, 2006).

The development of subunit vaccines has been based on the idea that an immune response engendered against selected immunodominant virus-encoded proteins would be sufficient to provide protection against infection or disease in a high-risk population. Most attention has focused on the immunodominant envelope glycoprotein gB (the product of the CMV *UL55* gene), and the pp65 tegument phosphoprotein (product of the *UL83* gene), the major CTL target in naturally CMV-seropositive people. A variety of protein expression strategies are currently in evaluation for these potential vaccine candidates: these include adjuvanted purified protein vaccines, virally-vectored vaccines and DNA vaccines.

Purified recombinant glycoprotein B vaccine: The protein subunit vaccine that has been studied most extensively to date is based on gB. The rationale for using gB is founded on the observation that this protein is the dominant target of virus-neutralizing antibody responses during natural infection (Britt, *et al.*, 1990). The gB formulation currently being used in vaccine trials is based on the CMV Towne strain sequence, and is expressed in Chinese hamster ovary (CHO) cells as a secreted protein. The protein was modified in two ways to facilitate its expression and purification. First, the proteolytic cleavage site, R-T-K-R – at which gB is normally cleaved – was mutated to allow the synthesis of uncleaved gB. Secondly, a stop mutation was introduced prior to its hydrophobic transmembrane domain, resulting in a truncated, soluble form of gB. The resulting secreted protein is purified from CHO cell culture supernatants and used, with adjuvant, as a CMV subunit vaccine candidate. Purified recombinant gB has been evaluated for safety and immunogenicity in several clinical trials, including vaccination in transplant patients.

Although most efforts in CMV subunit vaccine research have concentrated on gB, other envelope glycoproteins also elicit neutralizing antibody responses in the setting of natural infection. These include the gcII complex, consisting of glycoproteins gN (*UL73*) and gM (*UL100*), and the gcIII complex, consisting of glycoproteins gH (*UL75*), gO (*UL74*) and gL (*UL115*) (Kari and Gehrz, 1990; Mach, *et al.*, 2000; 2005). The gcII complex is of particular interest following the recent observation that this complex represents the most abundant glycoprotein in the viral envelope (Tullman, *et al.*, 2014).

Vectored subunit vaccines: In the vectored vaccine approach, the gene of interest is expressed in the context of a non-replicating (usually viral) vector. This approach holds the promise of inducing both cellular and humoral immune responses while maintaining a favorable safety profile (based on either the inability of the vector to complete its replication cycle in the vaccinee or the known safety of the vector in humans). Poxvirus vectors and alphavirus vectors have both been studied as potential CMV vaccine candidates. The poxvirus vector, which has received greatest emphasis for CMV vaccine development, is a canarypox vector known as ALVAC. ALVAC is attenuated vaccine strain of canarypox virus that replicates productively in avian species but abortively in mammalian cells. This feature provides an important safety barrier for human use and has facilitated the development of these vectors for various vaccine applications (Skinner, *et al.*, 2005). In addition, the ALVAC genome can accommodate large exogenous DNA fragments, providing great flexibility in the choice of antigen genes or gene combinations. Preclinical evaluation of an ALVAC-human CMV gB vaccine candidate in animals indicated that this recombinant induced strong humoral and CTL responses, justifying further development of this candidate for human clinical trials. Therefore, clinical trial evaluation has focused on using ALVAC-gB in a prime-boost approach, in which ALVAC vaccine is administered to prime immune responses for subsequent boost with live, attenuated vaccine, or recombinant protein vaccine. All vaccine regimens induced high-titer antibody and lymphoproliferative responses, but no benefit for priming or simultaneous vaccination was detected. Thus, ALVAC-gB priming resulted in augmented gB-specific responses following a boost with Towne vaccine, but not when followed by a subunit gB/MF59 boost.

Other poxvirus-based vectors have been developed for gB although they have not yet been tested in humans. A recombinant attenuated poxvirus was constructed – the modified vaccinia virus Ankara – that expresses a soluble, secreted form of CMV gB, based on the AD169 strain sequence (Wang, *et al.*, 2004). High gB-specific Nab levels were induced in immunized mice. Importantly, the presence of pre-existing poxvirus immunity in subjects did not appear to interfere with immune responses to subsequent immunizations with the gB construct, suggesting that this vaccine could be useful in those who have previously received the smallpox vaccine.

An important benefit of poxvirus-vectored vaccines may be their potential to stimulate CTL responses. To evaluate this possibility, ALVAC expressing pp65 was administered to CMV-seronegative adult volunteers in a placebo-controlled trial. The ALVAC/pp65 recipients developed CMV-specific CD8$^+$ CTL responses at frequencies comparable to those seen in naturally CMV-seropositive individuals. Recombinant vaccinia viruses expressing CMV targets have also shown encouraging signs of being tools for expansion of CMV-specific T-cells from CMV-seropositive donors. Recombinant vaccinia viruses and the related Ankara strain of vaccinia are capable of stimulating vigorous expansion of CMV-specific CD8$^+$ T-cells in CMV-positive donor peripheral blood mononuclear cells. These vectors, although unlikely to be used as primary CMV vaccines, may prove useful in generating large numbers of CMV-specific T-cells for adoptive immunotherapy applications. Ultimately, ALVAC-based vaccines that engender both CMV CTL and NAb responses, based on immunogens such as pp65 and gB, may merit clinical trial evaluation. Simultaneous administration of a glycoprotein immunogen (such as gB), and a CD8$^+$ T-cell target (such as pp65), may provide the best strategy for subunit-based vaccination approaches.

Another vectored vaccine approach is the use of attenuated alphavirus replicons (Lundstrom, 2003). Use of virus-like replicon particles based on the Venezuelan equine encephalitis (VEE) virus is especially attractive. Such particles express high levels of heterologous proteins, target expression to dendritic cells, and are capable of inducing both humoral and cellular immune responses to the vectored gene products. Using genetic approaches, VEE viruses can be generated that contain mutations in their envelope glycoproteins. Such mutations result in attenuated phenotypes, and foreign genes can be inserted in place of the VEE structural protein coding sequences. Using this approach, the CMV *gB*, *pp65* and *IE1* genes have been successfully cloned and expressed as VEE replicons, and preliminary immunogenicity studies in mice showed encouraging results.

<u>DNA vaccines:</u> DNA vaccines are based on the idea that cloned, immunogenic CMV gene products can induce protective immune responses when delivered in appropriate plasmid constructs. CMV DNA vaccines have been evaluated for immunogenicity in animal models and have been shown to induce both humoral and cellular immune responses. These studies have focused on gB and pp65 expression plasmids. In one study of human CMV gB DNA vaccines in mice, both the full-length gB, as well as a truncated, secreted form of gB (spanning amino acids 1–680 of 906 total gB residues), were evaluated. Immunization of mice with both constructs induced neutralizing antibodies, but titers were higher in mice immunized with the DNA encoding the truncated form of gB, lacking both its transmembrane and cytoplasmic domains. This study also examined mice immunized with a pp65-expressing plasmid. Both CTL and binding antibody responses were noted. When administered together (either at the same site, or at the same timepoint in different sites), the gB and pp65- expressing plasmids did not interfere with one another in their ability to elicit an immune response (Endresz, *et al.*, 1999). Subsequent studies indicated that immunogenicity could be augmented by inclusion of cytokines, aluminium phosphate or CpG oligodeoxynucleotides (Temperton, *et al.*, 2003).

CMV DNA vaccines have been evaluated in humans. Thus, phase I clinical trials have been conducted involving both a bivalent CMV DNA vaccine candidate, using plasmid DNA encoding pp65 and gB, and a trivalent vaccine candidate, which also includes a third plasmid encoding the IE1 gene product. The DNA vaccines being used in these clinical trials were formulated using the poloxamer adjuvant, CRL1005 and benzalkonium chloride and showed safe and well tolerated profile (Evans, *et al.*, 2004).

Several other CMV vaccine strategies are at various preclinical stages of development. One approach which looked promising in preclinical and animal model testing is the concept of peptide vaccination, using synthetic peptides comprising immunodominant CTL epitopes (Paston, Dodi and Madrigal, 2004). In one series of experiments, a pp65 human leucocyte antigen (HLA)-A2.1-restricted CTL epitope, corresponding to an immunodominant region spanning amino acid residues 495–503, was fused to the carboxyl terminus of a pan-DR T-help epitope. Preliminary work demonstrated the ability of this nonamer peptide to stimulate peripheral blood mononuclear cells from HLA A*0201 CMV-seropositive donors *in vitro*. Furthermore, the peptide was capable of eliciting CTL responses from mice transgenic for the same human HLA molecule (BenMohamed, *et al.*, 2000). DNA sequences corresponding to this epitope from

50 clinical CMV isolates indicated strong conservation of this sequence, suggesting that the vaccine could be broadly protective against multiple CMV strains. Clinical trials for this vaccine will focus on vaccinating bone marrow and stem cell donors, with the goal of transferring primed, CMV-specific CTL with the graft.

Another intriguing candidate for vaccination against CMV is the use of dense bodies (DBs) – enveloped, replication-defective particles, which are formed during replication of CMV in cell culture. A related particle, the noninfectious enveloped particle (NIEP), is also observed, but in smaller quantities than DBs, in tissue culture (Irmiere and Gibson, 1983). These structures contain the dominant target antigens for humoral (envelope glycoproteins) and cellular immune responses (pp65) elicited during natural infection. Upon their release from infected cultured cells, these particles can be purified by gradient centrifugation. DBs and NIEPs have been suggested as vaccine candidates, since although they contain a full repertoire of CMV proteins, they are noninfectious agents. DBs have been analysed for their ability to induce virus-neutralizing antibodies and CTL after immunization of mice (Pepperl-Klindworth, *et al.*, 2003). Based on these studies, DBs may represent a promising, novel approach to the development of a subunit vaccine against CMV infection.

In addition to programs examining strategies for active immunization against CMV disease, passive immunization has been employed for prevention or amelioration of CMV disease in high-risk individuals, *via* passive transfer of CMV-specific immunoglobulins or CMV-specific leucocytes. A successful application of the adoptive-transfer approach has been the demonstration of protective immunity to CMV following adoptive transfer of CMV-specific CTL to bone marrow and hemopoietic stem cell transplant recipients at high risk of CMV disease (Cobbold, *et al.*, 2001). The success of adoptive transfer in controlling CMV disease in this patient population supports the approach of pre-transplant vaccination of stem cell donors, in an effort to engender CMV-specific T-cells which can be adoptively transferred at the time of transplantation.

Another approach to passive immunization is the use of a CMV-specific immune globulin in high risk patients. Since many studies have been performed since the licensing of the antiviral drug ganciclovir, it is difficult to ascertain what is the protective effect of immunoglobulin alone in preventing CMV disease, since it is typically co-administered with antiviral therapy. Passive immunization has been demonstrated to reduce the occurrence of severe CMV disease in seronegative renal transplant recipients who received kidneys from CMV-seropositive donors (Snydman, *et al.*, 1987). Passive immunization has also been evaluated using monoclonal antibodies that target specific CMV proteins. An antibody that targeted the CMV gH protein was evaluated as an adjuvant therapy in transplant patients, but failed to show any benefit on antigenaemia, CMV disease or long-term survival (Boeckh, *et al.*, 2001). Another study of pregnant women with primary CMV infection examined the effect of intravenous CMV hyperimmune globulin (HIG) on congenital CMV infection and attendant sequelae and showed decreased CMV transmission and disability in newborns (Nigro, *et al.*, 2005). Vaccine candidates tested in humans are listed in Table **1**.

Table 1. Vaccine Candidates in Clinical Trials[a].

Vaccine Name	Type	Sponsor/ Manufacturer	Population	Efficacy
Towne	Live attenuated	None currently	Transplant recipients Children Women	Partial
Chimera	Towne/wild type recombinant	Aviron/ MedImmune	CMV-seropositive adults	TBD[b]
ALVAC	Canarypox pp65	Aventis Pasteur	CMV-seronegative adults	NT[c]
Subunit	Recombinant envelope (gB/MF59 or gB/alum)	Chiron	CMV-seronegative adults	50%[d]
ALVAC-Subunit	Canarypox followed by gB/MF59	Aventis Pasteur	CMV-seronegative adults	NT
Peptide fusion of CMV-CTL epitope	Peptide fusion of A2 and pp65 with helper peptide	City of Hope; Bachem	Transplant recipients	TBD
Particles	Dense bodies	Gutenberg University	CMV-seronegative adults	-
DNA	Plasmid	Vical	Transplant recipients	-

[a]Modified from Modlin, *et al.* (2004)
[b]TBD, To Be Determined
[c]NT, Not Tested
[d]Rieder F & Steininger (2014)

CURRENT STATE OF CMV VACCINES: AN UNMET NEED

Driven by the great public health importance of the problem of congenital CMV infection, CMV vaccines are moving forward in the clinic (Krause *et al.*, 2014). There is a need for an effective CMV vaccine that will protect immunocompromised transplant patients as well as newborns, although the key requirements for protection of these two populations (and the optimal vaccine strategy to provide this protection) may differ. To date, only the Towne vaccine – a live, attenuated CMV vaccine – has undergone efficacy evaluation. Unfortunately, there have been several pitfalls in the efficacy assessment in seronegative renal allograft recipients where it does not prevent infection but decreases frequency and mitigates severity of CMV-induced disease. In normal volunteers, it protects against parenteral low-dose wild-type challenge, while in mothers of children excreting CMV, it did not prevent infection. Moreover, it may be over-attenuated due to loss of certain genes (Quinnan, *et al.*, 1984). Application of molecular biological techniques, coupled with an improved understanding of the CMV genome, should allow design of safer, more immunogenic, live, attenuated vaccines. Adjuvanted protein vaccines, vectored vaccines and DNA vaccines are also undergoing evaluation in human volunteers. A CMV vaccine will need to induce both humoral responses and CTL responses.

CONFLICT OF INTEREST

The author confirms that this chapter contents have no conflict of interest.

ACKNOWLEDGEMENTS

Declared None.

ABBREVIATIONS

APC   = antigen-presenting cells

CMV   = cytomegalovirus

HHV   = human herpesvirus

Ig   = immunoglobulin

REFERENCES

Altman, J.D., Moss, P.A., Goulder, P.J., Barouch, D.H., McHeyzer-Williams, M.G., Bell, J.I., *et al.* 1996. Phenotypic analysis of antigen-specific T lymphocytes. *Science*, 274, 94–6.

Bancroft, G.J., Shellam, G.R. & Chalmer, J.E. 1981. Genetic influences on the augmentation of natural killer (NK) cells during murine cytomegalovirus infection: correlation with patterns of resistance. J Immunol, 126, 988–94.

BARRY, S.M., JOHNSON, M.A. & JANOSSY, G. 2000. Cytopathology or immunopathology? The puzzle of cytomegalovirus pneumonitis revisited. *Bone Marrow Transplant.*, 26, 591–7.

BenMohamed, L., Krishnan, R., Longmate, J., Auge, C., Low, L., Primus, J., *et al.* 2000. Induction of CTL response by a minimal epitope vaccine in HLA A*0201/DR1 transgenic mice: dependence on HLA class II restricted T(H) response. *Hum Immunol*, 61, 764–79.

Beutler, B., Crozat, K., Koziol, J.A. & Georgel, P. 2005. Genetic dissection of innate immunity to infection: the mouse cytomegalovirus model. Curr Opin Immunol, *17*, 36–43.

Boeckh, M., Bowden, R.A., Storer, B., Chao, N.J., Spielberger, R., Tierney, D.K., *et al.* 2001. Randomized, placebocontrolled, double-blind study of a cytomegalovirusspecific monoclonal antibody (MSL-109) for prevention of cytomegalovirus infection after allogeneic hematopoietic stem cell transplantation. *Biol Blood Marrow Transplant*, 7, 343–51.

Boeckh, M., Fries, B. & Nichols, W.G. 2004. Recent advances in the prevention of CMV infection and disease after hematopoietic stem cell transplantation. *Pediatr Transplant*, 8, 19–27.

Boppana, S.B., Ross, S.A., Novak, Z., Shimamura, M., Tolan, R.W. Jr, Palmer, A.L. *et al.* 2010. Dried blood spot real-time polymerase chain reaction assays to screen newborns for congenital cytomegalovirus infection. *JAMA*, 303, 1375-82.

Boutboul, F., Puthier, D., Appay, V., Pellé, O., Ait-Mohand, H., Combadière, B., *et al.* 2005. Modulation of interleukin-7 receptor expression characterizes differentiation of CD8 T cells specific for HIV, EBV and CMV. *AIDS*, 19, 1981–86.

Bowen, E.F., Sabin, C.A., Wilson, P., Griffiths, P.D., Davey, C.C., Johnson, M.A. *et al.* 1997. Cytomegalovirus (CMV) viraemia detected by polymerase chain reaction identifies a group of HIV-positive patients at high risk of CMV disease. *AIDS*, 11, 889-93.

Britt, W.J., Vugler, L., Butfiloski, E.J. & Stephens, E.B. 1990. Cell surface expression of human cytomegalovirus (HCMV) gp55-116 (gB): use of HCMVrecombinant vaccinia virusinfected cells in analysis of the human neutralizing antibody response. *J Virol*, 64, 1079–85.

Bukowski, J.F., Woda, B.A. & Welsh, R.M. 1984. Pathogenesis of murine cytomegalovirus infection in natural killer cell-depleted mice. J Virol, 52, 119–28.

Chalmer, J.E., Mackenzie, J.S. & Stanley, N.F. 1977. Resistance to murine cytomegalovirus linked to the major histocompatibility complex of the mouse. J Gen Virol, *37*, 107–14.

Chen, G., Shankar, P., Lange, C., Valdez, H., Skolnik, P.R., Wu, L., *et al.* 2001. CD8 T cells specific for human immunodeficiency virus, Epstein-Barr virus, and cytomegalovirus lack molecules for homing to lymphoid sites of infection. *Blood*, 98, 156–64.

Cobbold, M., Khan, N., Pourgheysari, B., Tauro, S., McDonald, D., Osman, H., *et al.* 2001. Adoptive transfer of cytomegalovirus-specific CTL to stem cell transplant patients after selection by HLA-peptide tetramers. *J Exp Med,* 202, 379–86.

Crozat, K., Georgel, P., Rutschmann, S., Mann, N., Du, X., Hoebe, K., *et al.* 2006. Analysis of the MCMV resistome by ENU mutagenesis. Mamm Genome, 17, 398–406.

Demmler, G.J. 1996. Congenital cytomegalovirus infection and disease. *Adv Pediatr Infect Dis*, 11, 135–62.

Elek, S.D. & Stern, H. 1974. Development of a vaccine against mental retardation caused by cytomegalovirus infection in utero. *Lancet*, 1, 1–5.

Endresz, V., Kari, L., Berencsi, K., Kari, C., Gyulai, Z., Jeney, C., *et al.* 1999. Induction of human cytomegalovirus (HCMV)- glycoprotein B (gB)-specific neutralizing antibody and phosphoprotein 65 (pp65)- specific cytotoxic T lymphocyte responses by naked DNA immunization. *Vaccine*, 17, 50–58.

Evans, T.G., Wloch, M., Hermanson, G., Selinsky, C., Geall, A. & Kaslow, D. 2004. Phase 1 trial of a bivalent, formulated plasmid DNA CMV vaccine for use in the transplant population (abstract). 44th Interscience Conference on Antimicrobial Agents and Chemotherapy (ICAAC). Washington DC.

Freeman, R.B. 2009. The 'indirect' effects of cytomegalovirus infection. *Am J Transplant,* 9, 11, 2453–8.

Fries, B.C., Chou, S., Boeckh, M. & Torok-Storb B. 1994. Frequency distribution of Cytomegalovirus envelope glycoprotein genotypes in bone marrow transplant recipients. *J Infect Dis*, 169, 769-74.

Gamadia, L.E., Remmerswaal, E.B., Weel, J.F., Bemelman, F., van Lier, R.A. & ten Berge, I.J. 2003. Primary immune responses to human CMV: a critical role for IFN-gamma-producing CD4+ T cells in protection against CMV disease. *Blood*, 101, 2686–92.

Gamadia, L.E., van Leeuwen, E.M., Remmerswaal, E.B., Yong, S.L., Surachno, S., Wertheim-van Dillen, P.M., *et al.* 2004. The size and phenotype of virus-specific T cell populations is determined by repetitive antigenic stimulation and environmental cytokines. *J Immunol*, 172, 6107–14.

Gonczol, E., Ianacone, J., Furlini, G., Ho, W. & Plotkin, S.A. 1989. Humoral immune response to cytomegalovirus Towne vaccine strain and to Toledo low-passage strain. *J Infect Dis*, 159, 851–9.

Griffiths, P.D. 2003. The indirect effects of virus infections. *Rev Med Virol*, 13, 1–3.

Huang, E.S., Huong, S.M., Tegtmeier, G.E. & Alford, C. 1980. Cytomegalovirus: genetic variation of viral genomes. *Ann N Y Acad Sci*, 354, 332-46.

Humar, A., Lebranchu, Y., Vincenti, F., Blumberg, E.A., Punch, J.D., Limaye, A.P. *et al.* 2010. The Efficacy and Safety of 200 Days Valganciclovir Cytomegalovirus Prophylaxis in High-Risk Kidney Transplant Recipients. *Am J Transplant*, 10, 1228-37.

Irmiere, A. & Gibson, W. 1983. Isolation and characterization of a noninfectious virion-like particle released from cells infected with human strains of cytomegalovirus. *Virology*, 130, 118–33.

Jacobson, M.A., Sinclair, E., Bredt, B., Agrillo, L., Black, D., Epling, C.L., *et al.* 2006. Safety and immunogenicity of Towne cytomegalovirus vaccine with or without adjuvant recombinant interleukin-12. *Vaccine*, 24, 5311-19.

Just, M., Buergin-Wolff, A., Emoedi, G. & Hernandez, R. 1975. Immunization trials with live attenuated cytomegalovirus TOWNE 125. *Infection*, 3, 111–14.

Kari, B. & Gehrz, R. 1990. Analysis of human antibody responses to human cytomegalovirus envelope glycoproteins found in two families of disulfide linked glycoprotein complexes designated gC-I and gC-II. *Arch Virol*, 114, 213–28.

Kern, F., Khatamzas, E., Surel, I., Frömmel, C., Reinke, P., Waldrop, S.L., *et al.* 1999. Distribution of human CMV-specific memory T cells among the CD8pos: subsets defined by CD57, CD27, and CD45 isoforms. *Eur J Immunol*, 29, 2908–15.

Kern, F., Bunde, T., Faulhaber, N., Kiecker, F., Khatamzas, E., Rudawski, I.M., *et al.* 2002. Cytomegalovirus (CMV) phophoprotein 65 makes a large contribution to shaping the T cell repertoire in CMV-exposed individuals. *J Infect Dis*, 185, 1709–16.

KRAUSE, P.R., BIALEK, S.R., BOPPANA, S.B., *ET AL.* 2014. Priorities for CMV vaccine development. *Vaccine*, 32, 4-10.

Kuijpers, T.W., Vossen, M.T., Gent, M.R., Davin, J.C., Roos, M.T., Wertheim-van Dillen, P.M., *et al.* 2003. Frequencies of circulating cytolytic, CD45RA+CD27–, CD8+ T lymphocytes depend on infection with CMV. *J Immunol*, 170, 4342–8.

Langeveld, M., Gamadia, L.E. & ten Berge, I.J.M. 2006. *T lymphocyte subset distribution in human spleen. Eur J Clin Invest*, 36, 250-6.

Lundstrom, K. 2003. Alphavirus vectors for vaccine production and gene therapy. *Expert Rev Vaccines*, 2, 447–59.

Mach, M., Kropff, B., Dal Monte, P. & Britt, W. 2000. Complex formation by human cytomegalovirus glycoproteins M (gpUL100) and N (gpUL73). *J Virol*, 74, 11881–92.

Mach, M., Kropff, B., Kryzaniak, M. & Britt, W. 2005. Complex formation by glycoproteins M and N of human cytomegalovirus: structural and functional aspects. *J Virol*, 79, 2160–70.

Miller, S., Seet, H., Khan, Y., Wright, C. & Nadarajah, R. 2010. Comparison of QIAGEN automated nucleic acid extraction methods for CMV quantitative PCR testing. *Am J Clin Pathol*, 133, 558-63.

Modlin, J.F., Arvin, A.M., Patricia Fast, P., Myers, M., Plotkin, S., & Rabinovich, R. 2004. Vaccine Development to Prevent Cytomegalovirus Disease: Report from the National Vaccine Advisory Committee. *Clin Infect Dis*, 39, 233-9.

Nigro, G., Adler, S.P., La Torre, R. & Best, A.M. 2005. Passive maternal immunization during pregnancy for congenital cytomegalovirus infection. *N Engl J Med*, 353, 1350–62.

Papagno, L., Appay, V., Sutton, J., Rostron, T., Gillespie, G.M., Ogg, G.S., *et al.* 2002. Comparison between HIV and CMV-specific T cell responses in long-term HIV infected donors. *Clin Exp Immunol*, 130, 509–17.

Paston, S.J., Dodi, I.A. & Madrigal, J.A. 2004. Progress made towards the development of a CMV peptide vaccine. *Hum Immunol*, 65, 544–9.

Pepperl-Klindworth, S., Frankenberg, N., Riegler, S. & Plachter, B. 2003. Protein delivery by subviral particles of human cytomegalovirus. *Gene Ther*, 10, 278–84.

Piñana, J.L., Martino, R., Barba, P., Margall, N., Roig, M.C., Valcárcel, D. *et al.* 2010. Cytomegalovirus infection and disease after reduced intensity conditioning allogeneic stem cell transplantation: single-centre experience. *Bone Marrow Transplant*, 45, 534-42.

PLOTKIN, S. 2000. The Prospects for Vaccination against Herpesviruses. The International Herpes Management Forum. http://www.ihmf.com/library/powerpoint/plotkin.ppt#5.

Porath, A., McNutt, R.A., Smiley, L.M. & Weigle, K.A. 1990. Effectiveness and cost benefit of a proposed live cytomegalovirus vaccine in the prevention of congenital disease. *Rev Infect Dis*, vol. 12, 31–40.

Price, P., Winter, J.G., Nikoletti, S., Hudson, J.B. & Shellam, G.R. 1987. Functional changes in murine macrophages infected with cytomegalovirus relate to H-2-determined sensitivity to infection. J Virol, 61, 3602–06.

Price, P., Gibbons, A.E. & Shellam, G.R. 1990. H-2 class I loci determine sensitivity to CMV in macrophages and fibroblasts. Immunogenetics, 32, 20–6.

Quinnan, G.V., Jr, Delery, M., Rook, A.H., Frederick, W.R., Epstein, J.S., Manischewitz, J.F., *et al.* 1984. Comparative virulence and immunogenicity of the Towne strain and a nonattenuated strain of cytomegalovirus. *Ann Intern Med*, 101, 478–83.

Rieder, F., Steininger, C. 2014. Cytomegalovirus vaccine: phase II clinical trial results. *Clin Microbiol Infect*, 20 (Suppl 5), 95-102.

Sandberg, J.K., Fast, N.M. & Nixon, D.F. 2001. Functional heterogeneity of cytokines and cytolytic effector molecules in human CD8+ T lymphocytes. *J Immunol*, 167, 181–7.

Scalzo, A.A., Corbett, A.J., Rawlinson, W.D., Scott, G.M. & Degli-Esposti, M.A. 2007. The interplay between host and viral factors in shaping the outcome of cytomegalovirus infection. *Immunol Cell Biol*, 85, 46–54.

Schleiss, M. 2005. Progress in Cytomegalovirus Vaccine Development. *Herpes*, 12, 66-75.

Schleiss, M. 2008. Cytomegalovirus Vaccine Development. *Curr Top Microbiol Immunol*, 325, 361–82.

Skinner, M.A., Laidlaw, S.M., Eldaghayes, I., Kaiser, P. & Cottingham, M.G. 2005. Fowlpox virus as a recombinant vaccine vector for use in mammals and poultry. *Expert Rev Vaccines*, 4, 63–76.

Smith, S.H., Brown, M.H., Rowe, D., Callard, R.E. & Beverley, P.C. 1986. Functional subsets of human helper-inducer cells defined by a new monoclonal antibody, UCHL1. *Immunology*, 58, 63–70.

Snydman, D.R., Werner, B.G., Heinze-Lacey, B., Berardi, V.P., Tilney, N.L., Kirkman, R.L., *et al.* 1987. Use of cytomegalovirus immune globulin to prevent cytomegalovirus disease in renal-transplant recipients. *N Engl J Med*, 317, 1049–54.

Spector, S. *et al.* 2006. GMHC Treatment Issues.

Stanberry, L.R., Rosenthal, S.L., Mills, L., Succop, P.A., Biro, F.M., Morrow, R.A. *et al.* 2004. Longitudinal risk of herpes simplex virus (HSV) type 1, HSV type 2, and cytomegalovirus infections among young adolescent girls. *Clin Infect Dis*, 39, 1433–8.

Starr, S.E., Glazer, J.P., Friedman, H.M., Farquhar, J.D. & Plotkin, S.A. 1981. Specific cellular and humoral immunity after immunization with live Towne strain cytomegalovirus vaccine. *J Infect Dis*, 143, 585–9.

Strangert, K., Carlstrom, G., Jeansson, S. & Nord, C.E. 1976. Infections in preschool children in group day care. *Acta Paediatr Scand*, 65, 455–63.

Sylwester, A.W., Mitchell, B.L., Edgar, J.B., Taormina, C., Pelte, C., Ruchti, F., *et al.* 2005. Broadly targeted human cytomegalovirus-specific CD4+ and CD8+ T cells dominate the memory compartments of exposed subjects. *J Exp Med*, 202, 673–85.

Temperton, N.J., Quenelle, D.C., Lawson, K.M., Zuckerman, J.N., Kern, E.R., Griffiths, P.D., *et al.* 2003. Enhancement of humoral immune responses to a human cytomegalovirus DNA vaccine: adjuvant effects of aluminum phosphate and CpG oligodeoxynucleotides. *J Med Virol*, 70, 86–90.

Tu, W., Chen, S., Sharp, M., Dekker, C., Manganello, A.M., Tongson, E.C., *et al*. 2004. Persistent and selective deficiency of CD4$^+$ T cell immunity to cytomegalovirus in immunocompetent young children. *J Immunol*, 172, 3260–7.

Tullman, j.a., Mary-Elizabeth Harmon, m.-e., Michael Delannoy, m. & gibson, w. 2014. Recovery of an HMWP/hmwBP (pUL48/pUL47) Complex from Virions of Human Cytomegalovirus: Subunit Interactions, Oligomer Composition, and Deubiquitylase Activity. *J Virol*, 88(15), 8256-67.

van Leeuwen, E.M.M., de Bree, G.J., ten Berge, I.J.M. & van Lier, R.A.W. 2006. Human virus-specific CD8+ T cells: diversity specialists. *Immunol Rev*, 211, 225–35.

Wang, Z., La Rosa, C., Maas, R., Ly, H., Brewer, J., Mekhoubad, S., *et al*. 2004. Recombinant modified vaccinia virus Ankara expressing a soluble form of glycoprotein B causes durable immunity and neutralizing antibodies against multiple strains of human cytomegalovirus. *J Virol*, 78, 3965–76.

Welsh, R.M., Selin, L.K. & Szomolanyi-Tsuda. E. 2004. Immunological memory to viral infections. *Annu Rev Immunol*, 22, 711–43.

Wherry, E.J., Barber, D.L., Kaech, S.M., Blattman, J.N. & Ahmed, R. 2004. Antigen-independent memory CD8 T cells do not develop during chronic viral infection. *Proc Natl Acad Sci USA*, 101, 16004–9.

Whitley, R.J. 2004. Congenital cytomegalovirus infection: epidemiology and treatment. *Adv Exp Med Biol*, 549, 155–60.

Wykes, M.N., Shellam, G.R., McCluskey, J., Kast, W.M., Dallas, P.B. & Price, P. 1993. Murine cytomegalovirus interacts with major histocompatibility complex class I molecules to establish cellular infection. J Virol, 67, 4182–89.

Varicella-Zoster Virus Infections and Vaccine Advances

Kallie Appleton[1], Ghaith Al Eyd[2] and Liljana Stevceva[2,*]

[1]The Paul L. Foster School of Medicine, El Paso, TX, USA; [2]California Northstate University College of Medicine, Elk Grove, CA, USA

Abstract: Varicella Zoster Virus (VZV) causes a highly contagious infection, chicken pox and has the capacity to cause latent persistent infection of the dorsal root ganglia. The latent virus can reemerge inducing painful blistering of the skin called herpes zoster or shingles. Worldwide varicella immunization with an attenuated OKA/Merck VZV considerably reduced the incidence of the disease but did not prevent latency. Introduction of a second dose of the vaccine and the vaccine for people over 50, Zostavax, consisting of a much larger dose of the same vaccine virus, reduced the incidence of chicken pox but did not prevent latency and reemergence of the virus as herpes zoster. This review provides an overview of the structure of the virus, mechanism of infection and known evasion mechanisms that VZV uses to establish latency and persistence.

Keywords: Varicella-Zoster, VZV, Herpesviridae, Aplhaherpesviridae, Chicken pox, dorsal root ganglia, cranial root ganglia, Varicella vaccine, Viral vaccine, Varivax, Zostavax, Vaccination, Virus disease, Herpesvirus 3, Herpes zoster, HSV, Shingles, Immunogenicity

INTRODUCTION TO VARICELLA ZOSTER VIRUS

VZV is classified as part of the *Herpesviridae* family, the *Alphaherpesvirinae* subfamily, and the *Varicellovirus* genus (Cohen *et al.* 2006, as reviewed in (Yamagishi *et al.*, 2008)). It is important to understand the biological aspects of the virus, identify the clinical presentation of the viral infection, and apply this knowledge to the modern treatment and prevention of VZV infection.

Before the introduction of varicella vaccine, a nationwide study of 21,288 participants from US was performed and showed that 96.3% of the study participants had been exposed to the VZV as determined by the presence of VZV-

*****Corresponding author Liljana Stevceva:** University of Texas Rio Grande Valley School of Medicine, 2102 Treasure Hills Blvd., Harlingen TX, USA; E-mail: Liljana@hotmail.com

specific IgG antibodies. There was an 86% prevalence of these antibodies in children aged 6-11 years and 99.6% prevalence in persons 40-49 years of age. Persons older than 49 years also showed a prevalence of 99.6% or even higher. These findings show that a large majority of the human population in the USA has been exposed to VZV (2).

VZV particularly infects T cells, cutaneous epithelial cells and dorsal root ganglia cells. The infected T cells express skin homing proteins and other activation markers *e.g.* cutaneous leukocyte antigen (CLA) and they tend to circulate through the skin and other tissues. The infected T cells deliver VZV to the skin cells and contribute to the gradual formation of skin lesions. Thus, T cell tropism appears to be necessary for primary VZV infection (Zerboni *et al.*, 2014). There are two common clinical presentations of VZV infection, varicella (chicken pox) and herpes zoster (shingles) (reviewed in (Goldman L., 2012), (Arvin, 1996)). Primary VZV infection presents clinically as chicken pox, a highly contagious febrile disease accompanied by a characteristic generalized pruritic skin rash. The rash appears first on the head, chest and the back as maculopapular rash and later progresses to vesicular before crusting. The lesions spread from the head to the rest of the body, but are mostly concentrated on the chest and the back. Chicken pox rash develops in crops so, lesions at different stage of the process can be seen simultaneously. Vesicles may also be seen on mucosae (enanthema) (reviewed in (Goldman L., 2012), (Arvin, 1996).

After the initial infection, VZV remains dormant in the cranial nerve ganglia and dorsal root ganglia (reviewed in (Goldman L., 2012). This latent virus may reemerge causing shingles, which is a painful rash occurring in a unilateral pattern on a particular dermatome (reviewed in (Arvin, 1996).

Other, less common clinical presentations of VZV infection include involvement of large, or small cerebral arteries (VZV vasculopathy) that can present as fever, headache, altered mental status, and ischemic attacks including focal deficits caused by stroke. VZV myelopathy is another rare clinical presentation of VZV infection that presents as self-limiting, post infectious myelitis in immunocompetent patients resulting in a spastic paraparesis with sphincter problems and sometimes, sensory changes. In immunocompromised patients, the VZV myelopathy can be progressive and sometimes even fatal. Eye involvement is manifested as either acute retinal necrosis in both immunocompetent and immunocompromised patients or progressive outer retinal necrosis that occurs commonly as an opportunistic retinal infection in AIDS patients in USA. Reactivation of VZV can also cause pain without rash in a condition called *zoster sine herpete* ([reviewed in (Mueller, 2008)].

The primary VZV infection induces life-long immunity against newly acquired VZV infections. This has been used to develop and introduce a preventative vaccine consisting of a live, attenuated varicella virus. The varicella vaccine (Varivax, Merck and Company, Inc, West Point, PA) was licensed on March 17, 1995 by the US Food and Drug Administration (FDA) for use in persons who have not had chicken pox and are aged 12 months or older.

STRUCTURE & GENOME OF THE VIRUS

The size of VZV ranges from 150 to 200 nm diameter (Mueller, 2008). VZV contains a linear double-stranded DNA. A study that mapped out the entire genome of VZV reported that VZV contains around 124,884 base pairs of nucleotides, approximately 69-71 open reading frames and has a long and a short segment (reviewed in (Arvin, 1996), (Mueller, 2008), (Davison and Scott, 1986). It has also been noted that VZV has genomic similarities to Herpes Simplex-1 Virus (HSV-1) (Yamagishi *et al.*, 2008).

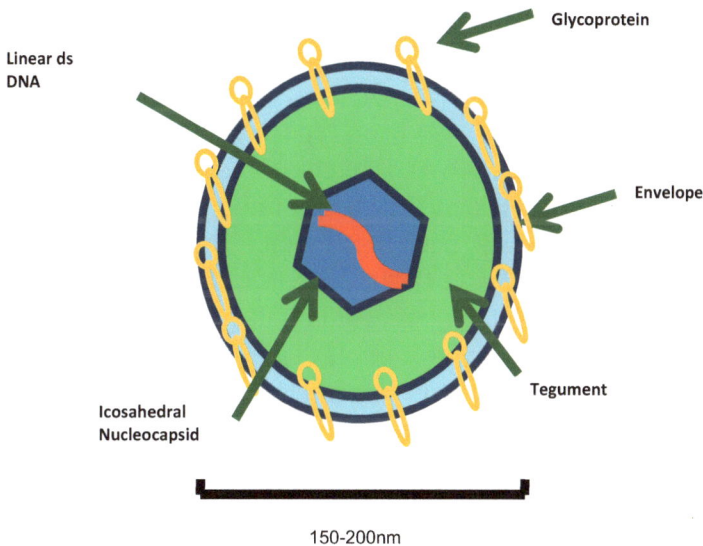

Fig. (1). The *Varicella Zoster* virus structure: includes protein tegument, icosahedral-shaped nucleocapsid, & envelope glycoproteins

Surrounding this linear double-stranded DNA is a nucleocapsid that is icosahderal (Mueller, 2008). There is a protein tegument around the nucelocapsid (Cohen *et al.* 1995, as reviewed in (Arvin, 1996). The purpose of the tegument, similarly to the one in *Herpesvirus*, is to give the virus integrity and structural support, and to help the virus release its newly created capsids (Mettenleiter, 2002), (Tischer *et al.*, 2007). Finally, there is a lipid envelope around the VZV tegument (Arvin, 1996), (Mueller, 2008). A schematic representation of the viral structure is seen in Fig. (**1**).

Glycoproteins

It has been discovered over the years that VZV envelope contains many glycoproteins (Grose, 1980), (Grose *et al.*, 1984), (Grose and Friedrichs, 1982). VZV glycoproteins include gB, gC, gE, gH, gI, gK, and gL, as well as possibly gM and gN (Cohen *et al.* 2006) (Yamagishi *et al.*, 2008).

Glycoproteins B and C (gB, gC) are sites where host antibodies can bind and neutralize VZV infectivity (these glycoproteins encode polypeptides with neutralization epitopes). It has been demonstrated that anti-gB monoclonal antibodies can neutralize VZV infectivity in the absence of complement while anti-gC monoclonal antibodies can only neutralize VZV in the presence of complement. (Keller *et al.*, 1984). Glycoprotein B may also play a role in the VZV entry into the host cell (Arvin, 1996). Glycoprotein E is the most abundant viral glycoprotein found in VZV-infected cells (Arvin, 1996).

The glycoproteins of VZV have also been reported to interact with each other in different ways (Forghani *et al.*, 1994), (Hutchinson *et al.*, 1992). Glycoproteins L and H of VZV form a complex in a fashion similar to the gH and gL of HSV where these two glycoproteins need each other to be able to perform their functions and operations (13), (14). It has been found that gH requires the presence of gL to be appropriately expressed on the surface of HSV-infected cells (Hutchinson *et al.*, 1992). Herpes simplex gH-gL complex activates gB to promote cell-cell fusion and viral skin tropism by triggering its fusogenic function (Zerboni *et al.*, 2014). Prompted by the findings of these studies, another study looked at the relationship of gL and gH in VZV, and found that these two glycoproteins formed a complex in this virus as well (Forghani *et al.*, 1994), (Hutchinson *et al.*, 1992).

Similarly, it has been shown that gI and gE form a complex (Yao *et al.*, 1993), and that the N- terminus of gI is important in the formation and stability of the bond of gI-gE complex (Kimura *et al.*, 1997). In fact, scientists have proven that without gI-gE complex, gI could not even make it to the surface of the VZV particle (Oliver *et al.*, 2009). When evaluating the ability of the virus to infect skin cells, it has been concluded that the gI-gE complex was necessary for successful virulence of the virus (Oliver *et al.*, 2009). GE binds to the Fc portion of immunoglobulin G as shown in Fig. (**2**) ([Kapsenberk 1964, as reviewed in (Arvin, 1996)]. GI has also been shown to contribute to the virulence of the virus outside of its role in complexing with gE (Oliver *et al.*, 2009).

Glycoproteins in VZV participate in viral-cell membrane fusion. The gH-gL complex in VZV has been shown to play a role in fusion of the viral membrane to the cell membrane during infection (Duus *et al.*, 1995). This was confirmed by Maresova *et al.* with an elegant set of experiments using transient expression system with Vaccinia Virus where co-expression of VZV gH and gL induced formation of a large syntytium in absence of any other VZV glycoproteins. Expression of VZV gB alone induced small syntytium confirming the role of gB in the fusion of host cell membrane to viral membrane. When gE was co-expressed with gB a larger syncytium was seen than with gB alone (Maresova *et al.*, 2001).

Glycoprotein I has been proven to play a pivotal role in the growth of VZV. In a study done by Cohen *et al.* in 1997, mutant VZV, deleted in gI was created and used to infect human, guinea pig and monkey cells. The growth curves of the mutant virus were stunted in nonhuman cells but not as much in human melanoma, schwannoma and fibroblast cells (Cohen and Nguyen, 1997).

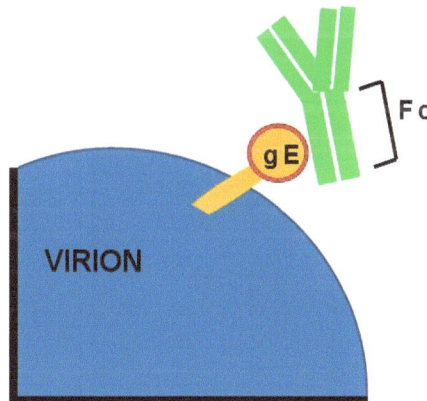

Fig. (2). Glycoprotein E, in the envelope of the virion, binds to the Fc region of IgG.

Glycoprotein I (gI) is important for infection of human dorsal root ganglia (DRGs) *in vivo*. In a set of experiments, gI was deleted in the OKA VZV strain and the role of gI in VZV neuropathogenesis was examined by inoculating dorsal root ganglia (DRG) xenografts with a gI deletion mutant. Without gI, VZV genome copies were 100-fold lower at 14 days after inoculation; only 9% of DRGs yielded infectious virus compared with 67% of those infected with intact VZV, and only 5–10% of cells within DRGs were infected. While intact recombinant Oka (rOka) initiated a short replicative phase followed by persistence in DRGs, the gI deletion mutant, rOka_gI, showed prolonged replication with no transition to persistence up to 70 days after infection. This indicates that gI plays an important role in establishment of viral persistence in dorsal root ganglia (Zerboni *et al.*, 2007). In a subsequent set of experiments it has been shown that gI is required for axonal localization of the virus and it might be facilitating axonal spread (Christensen *et al.*, 2013)

VZV is transmitted through aerosol droplets and cutaneous vesicle fluid. VZV envelope proteins are known to be able to bind mannose receptors on immature dendritic cells. Thus, following exposure, VZV infects the dendritic cells of the nasal pharyngeal mucosa and Langerhans and plasmacytoid dendritic cells of the respiratory mucosa. After inoculation there is an incubation period where the virus is undetectable for about 10-21 days during which time it is believed that the virus spreads to the lymph nodes. VZV-infected dendritic cells show no significant decrease in cell viability or evidence of apoptosis and do not exhibit altered cell surface levels of major histocompatibility complex (MHC) class I, MHC class II, CD86, CD40, or CD1a (Abendroth *et al.*, 2001). Infected dendritic cells travel to regional lymph nodes and have the ability to efficiently transfer the infection to both CD4+ and CD8+ T lymphocytes. These VZV infected T cells produce infectious virus and can infect fibroblasts and other cells. Infected T cells are resistant to cell death but require cell-to-cell contact to infect other cells as cell-free infectious VZV is not produced (Abendroth *et al.*, 2001). It is proposed that the infected T cells home to the skin and then infect surrounding cells causing the typical vesicular lesions of chicken pox (Ku *et al.*, 2004). Infection of dorsal ganglia occurs mostly through retrograde infection of the axons and also through hematogenous transmission (Eshleman *et al.*, 2011).

Varicella vaccine consists of live, attenuated OKA/Merck Varicella virus (Quinlivan *et al.*, 2011) (Weibel *et al.*, 1984). The vaccine was developed by Merk, Sharp & Dohme in the 1980 from the OKA strain virus, isolated and attenuated by Takahashi *et al.* in Japan (Takahashi *et al.*, 1974), (Takahashi *et al.*, 1975). As previously mentioned, the FDA approved the vaccine in 1995. The American Academy of Pediatrics and the Advisory Committee for Immunization Practices have recommended the use of the vaccine in 1995, 1996, and 1999 respectively (American Academy of Pediatrics, 2000). Two doses of the vaccine applied subcutaneously induce life-long protection against chicken pox and had reduced its incidence considerably. For example, the successful Varicella vaccination reduced overall incidence of chicken pox in all age groups in Massachusetts for more than 66%. At the same time, incidence of herpes zoster increased from 2.77/1000 in 1999 to 5.25/1000 raising speculations that the two might be connected (Yih *et al.*, 2005). Similar findings were reported in California (Civen *et al.*, 2009), Australia (Carville *et al.*, 2010), (Kelly *et al.*, 2014) and Taiwan (Chao *et al.*, 2012). Although the attenuated vaccine virus strain is much less neurovirulent, there have been reports of zoster appearing, usually in the vaccinated arm, three or more years after vaccination indicating that the vaccine strain still has the capacity to establish latent infection in the cervical dorsal root ganglia (Horien and Grose, 2012). Still, reports on herpes zoster incidence in vaccinated population are controversial and more studies are needed to draw sound conclusions on this issue (Marin *et al.*, 2008). An US study that has examined the link between herpes zoster incidence and childhood varicella vaccination concluded that the increase in the incidence of herpes zoster did not seem to be related to the varicella vaccine as age specific increase in herpes zoster in the US has occurred even before the introduction of the vaccine (Hales *et al.*, 2013). Another study done in Alberta, Canada showed that after eight years of implementing varicella vaccination program, the rates of medically attended shingles for children below age of 10 years decreased. The increased rates of shingles in older persons have been observed before implementing the vaccine program and was thus considered not to be related to the vaccination (Russell *et al.*, 2014). Despite these reports, the concern about increasing incidence of herpes zoster with age prompted FDA approval in 2006 of Zostavax, vaccine targeted to adults and the subsequent extension of its application in 2011 to 50-59 years old. Zostavax contains the same attenuated VZV but it is given at a much higher dose of 19,400 PFU in the 0.65ml dose compared to 1350 PFU in 0.5ml dose of varicella vaccine. Zostavax unfortunately does not prevent occurrence of herpes zoster but only reduces its incidence for 51.3% (Oxman *et al.*, 2005).

Immunity acquired by immunization with the attenuated varicella vaccine is mediated by induction of neutralizing antibodies against the virus. FDA established that gpELISA titer of >5 units/ml is an approximate correlate of protection and chances of breakthrough chicken pox infection over 7 years after immunization are 3.5 times higher if the titer is lower than 5 units/ml (Krause and Klinman, 1995). The titer also correlates with the detection of VZV-specific proliferative T cell responses although adults require additional dose to achieve seroconversion (Nader *et al.*, 1995). In this study, VZV-specific cell mediated immune responses were significantly lower in adults at one year after immunization despite receiving two doses of the vaccine compared to only one dose in the children. This was the first indication that the limitation of the adult helper T cell response might signal increased susceptibility for infection and a need for a boost in adults in order to maintain the protection against the virus.

In summary, although an effective varicella vaccine has been in use for a long time and Zostavax truly lowers the incidence of herpes zoster, neither of these vaccines in essence prevents the VZV going into latency or persistence. In fact, to date there is no vaccine candidate that specifically targets virus evasion mechanisms in order to develop a vaccine that will prevent the VZV from establishing latent and persistent infection and/or serve as a therapeutic vaccine that will help the body immune system to overcome the virus evasion mechanisms.

CONFLICT OF INTEREST

The authors confirm that this chapter contents have no conflict of interest.

ACKNOWLEDGEMENTS

Declared None.

REFERENCES

ABENDROTH, A., MORROW, G., CUNNINGHAM, A. L. & SLOBEDMAN, B. 2001. Varicella-zoster virus infection of human dendritic cells and transmission to T cells: implications for virus dissemination in the host. *J Virol,* 75, 6183-92.

AMERICAN ACADEMY OF PEDIATRICS, C. O. I. D. 2000. Varicella Updates. *Pediatrics,* 105, 136-141.

ARVIN, A. M. 1996. Varicella-Zoster Virus. *Clinical Microbiology Reviews,* 9, 361-381.

CARVILLE, K. S., RIDDELL, M. A. & KELLY, H. A. 2010. A decline in varicella but an uncertain impact on zoster following varicella vaccination in Victoria, Australia. *Vaccine,* 28, 2532-8.

CHAO, D. Y., CHIEN, Y. Z., YEH, Y. P., HSU, P. S. & LIAN, I. B. 2012. The incidence of varicella and herpes zoster in Taiwan during a period of increasing varicella vaccine coverage, 2000-2008. *Epidemiol Infect,* 140, 1131-40.

CHRISTENSEN, J., STEAIN, M., SLOBEDMAN, B. & ABENDROTH, A. 2013. Varicella-zoster virus glycoprotein I is essential for spread in dorsal root ganglia and facilitates axonal localization of structural virion components in neuronal cultures. *J Virol,* 87, 13719-28.

CIVEN, R., CHAVES, S. S., JUMAAN, A., WU, H., MASCOLA, L., GARGIULLO, P. & SEWARD, J. F. 2009. The incidence and clinical characteristics of herpes zoster among children and adolescents after implementation of varicella vaccination. *Pediatr Infect Dis J,* 28, 954-9.

COHEN, J. I. & NGUYEN, H. 1997. Varicella-zoster virus glycoprotein I is essential for growth of virus in Vero cells. *J Virol,* 71, 6913-20.

DAVISON, A. J. & SCOTT, J. E. 1986. The complete DNA sequence of varicella-zoster virus. *J Gen Virol,* 67 (Pt 9), 1759-816.

DUUS, K. M., HATFIELD, C. & GROSE, C. 1995. Cell surface expression and fusion by the varicella-zoster virus gH:gL glycoprotein complex: analysis by laser scanning confocal microscopy. *Virology,* 210, 429-40.

ESHLEMAN, E., SHAHZAD, A. & COHRS, R. J. 2011. Varicella zoster virus latency. *Future Virol,* 6, 341-355.

FORGHANI, B., NI, L. & GROSE, C. 1994. Neutralization epitope of the varicella-zoster virus gH:gL glycoprotein complex. *Virology,* 199, 458-62.

GOLDMAN L., S. A. 2012. Varicella-Zoster Virus (Chickenpox, Shingles). *Goldman's Cecil Medicine.* Philadelphia: Elesevier.

GROSE, C. 1980. The synthesis of glycoproteins in human melanoma cells infected with varicella-zoster virus. *Virology,* 101, 1-9.

GROSE, C., EDWARDS, D. P., WEIGLE, K. A., FRIEDRICHS, W. E. & MCGUIRE, W. L. 1984. Varicella-zoster virus-specific gp140: a highly immunogenic and disulfide-linked structural glycoprotein. *Virology,* 132, 138-46.

GROSE, C. & FRIEDRICHS, W. E. 1982. Immunoprecipitable polypeptides specified by varicella-zoster virus. *Virology,* 118, 86-95.

HALES, C. M., HARPAZ, R., JOESOEF, M. R. & BIALEK, S. R. 2013. Examination of links between herpes zoster incidence and childhood varicella vaccination. *Ann Intern Med,* 159, 739-45.

HORIEN, C. & GROSE, C. 2012. Neurovirulence of varicella and the live attenuated varicella vaccine virus. *Semin Pediatr Neurol,* 19, 124-9.

HUTCHINSON, L., BROWNE, H., WARGENT, V., DAVIS-POYNTER, N., PRIMORAC, S., GOLDSMITH, K., MINSON, A. C. & JOHNSON, D. C. 1992. A novel herpes simplex virus glycoprotein, gL, forms a complex with glycoprotein H (gH) and affects normal folding and surface expression of gH. *J Virol,* 66, 2240-50.

KELLER, P. M., NEFF, B. J. & ELLIS, R. W. 1984. Three major glycoprotein genes of varicella-zoster virus whose products have neutralization epitopes. *J Virol,* 52, 293-7.

KELLY, H. A., GRANT, K. A., GIDDING, H. & CARVILLE, K. S. 2014. Decreased varicella and increased herpes zoster incidence at a sentinel medical deputising service in a setting of increasing varicella vaccine coverage in Victoria, Australia, 1998 to 2012. *Euro Surveill,* 19.

KIMURA, H., STRAUS, S. E. & WILLIAMS, R. K. 1997. Varicella-zoster virus glycoproteins E and I expressed in insect cells form a heterodimer that requires the N-terminal domain of glycoprotein I. *Virology,* 233, 382-91.

KRAUSE, P. R. & KLINMAN, D. M. 1995. Efficacy, immunogenicity, safety, and use of live attenuated chickenpox vaccine. *J Pediatr,* 127, 518-25.

KU, C. C., ZERBONI, L., ITO, H., GRAHAM, B. S., WALLACE, M. & ARVIN, A. M. 2004. Varicella-zoster virus transfer to skin by T Cells and modulation of viral replication by epidermal cell interferon-alpha. *J Exp Med,* 200, 917-25.

MARESOVA, L., PASIEKA, T. J. & GROSE, C. 2001. Varicella-zoster Virus gB and gE coexpression, but not gB or gE alone, leads to abundant fusion and syncytium formation equivalent to those from gH and gL coexpression. *J Virol,* 75, 9483-92.

MARIN, M., MEISSNER, H. C. & SEWARD, J. F. 2008. Varicella prevention in the United States: a review of successes and challenges. *Pediatrics,* 122, e744-51.

METTENLEITER, T. C. 2002. Herpesvirus assembly and egress. *J Virol,* 76, 1537-47.

MUELLER, N. H. 2008. VZV Infection: Clinical Features and Molecular Pathogenesis of disease and Latency. *Neurologic Clinics,* 26.

NADER, S., BERGEN, R., SHARP, M. & ARVIN, A. M. 1995. Age-related differences in cell-mediated immunity to varicella-zoster virus among children and adults immunized with live attenuated varicella vaccine. *J Infect Dis,* 171, 13-7.

OLIVER, S. L., SOMMER, M., ZERBONI, L., RAJAMANI, J., GROSE, C. & ARVIN, A. M. 2009. Mutagenesis of varicella-zoster virus glycoprotein B: putative fusion loop residues are essential for viral replication, and the furin cleavage motif contributes to pathogenesis in skin tissue *in vivo. J Virol,* 83, 7495-506.

OXMAN, M. N., LEVIN, M. J., JOHNSON, G. R., SCHMADER, K. E., STRAUS, S. E., GELB, L. D., ARBEIT, R. D., SIMBERKOFF, M. S., GERSHON, A. A., DAVIS, L. E., WEINBERG, A., BOARDMAN, K. D., WILLIAMS, H. M., ZHANG, J. H., PEDUZZI, P. N., BEISEL, C. E.,

MORRISON, V. A., GUATELLI, J. C., BROOKS, P. A., KAUFFMAN, C. A., PACHUCKI, C. T., NEUZIL, K. M., BETTS, R. F., WRIGHT, P. F., GRIFFIN, M. R., BRUNELL, P., SOTO, N. E., MARQUES, A. R., KEAY, S. K., GOODMAN, R. P., COTTON, D. J., GNANN, J. W., JR., LOUTIT, J., HOLODNIY, M., KEITEL, W. A., CRAWFORD, G. E., YEH, S. S., LOBO, Z., TONEY, J. F., GREENBERG, R. N., KELLER, P. M., HARBECKE, R., HAYWARD, A. R., IRWIN, M. R., KYRIAKIDES, T. C., CHAN, C. Y., CHAN, I. S., WANG, W. W., ANNUNZIATO, P. W., SILBER, J. L. & SHINGLES PREVENTION STUDY, G. 2005. A vaccine to prevent herpes zoster and postherpetic neuralgia in older adults. *N Engl J Med,* 352, 2271-84.

QUINLIVAN, M., BREUER, J. & SCHMID, D. S. 2011. Molecular studies of the Oka varicella vaccine. *Expert Rev Vaccines,* 10, 1321-36.

RUSSELL, M. L., DOVER, D. C., SIMMONDS, K. A. & SVENSON, L. W. 2014. Shingles in Alberta: before and after publicly funded varicella vaccination. *Vaccine,* 32, 6319-24.

TAKAHASHI, M., OKUNO, Y., OTSUKA, T., OSAME, J. & TAKAMIZAWA, A. 1975. Development of a live attenuated varicella vaccine. *Biken J,* 18, 25-33.

TAKAHASHI, M., OTSUKA, T., OKUNO, Y., ASANO, Y. & YAZAKI, T. 1974. Live vaccine used to prevent the spread of varicella in children in hospital. *Lancet,* 2, 1288-90.

TISCHER, B. K., KAUFER, B. B., SOMMER, M., WUSSOW, F., ARVIN, A. M. & OSTERRIEDER, N. 2007. A self-excisable infectious bacterial artificial chromosome clone of varicella-zoster virus allows analysis of the essential tegument protein encoded by ORF9. *J Virol,* 81, 13200-8.

WEIBEL, R. E., NEFF, B. J., KUTER, B. J., GUESS, H. A., ROTHENBERGER, C. A., FITZGERALD, A. J., CONNOR, K. A., MCLEAN, A. A., HILLEMAN, M. R., BUYNAK, E. B. & ET AL. 1984. Live attenuated varicella virus vaccine. Efficacy trial in healthy children. *N Engl J Med,* 310, 1409-15.

YAMAGISHI, Y., SADAOKA, T., YOSHII, H., SOMBOONTHUM, P., IMAZAWA, T., NAGAIKE, K., OZONO, K., YAMANISHI, K. & MORI, Y. 2008. Varicella-zoster virus glycoprotein M homolog is glycosylated, is expressed on the viral envelope, and functions in virus cell-to-cell spread. *J Virol,* 82, 795-804.

YAO, Z., JACKSON, W., FORGHANI, B. & GROSE, C. 1993. Varicella-zoster virus glycoprotein gpI/gpIV receptor: expression, complex formation, and antigenicity within the vaccinia virus-T7 RNA polymerase transfection system. *J Virol,* 67, 305-14.

YIH, W. K., BROOKS, D. R., LETT, S. M., JUMAAN, A. O., ZHANG, Z., CLEMENTS, K. M. & SEWARD, J. F. 2005. The incidence of varicella and herpes zoster in Massachusetts as measured by the Behavioral Risk Factor Surveillance System (BRFSS) during a period of increasing varicella vaccine coverage, 1998-2003. *BMC Public Health,* 5, 68.

ZERBONI, L., REICHELT, M., JONES, C. D., ZEHNDER, J. L., ITO, H. & ARVIN, A. M. 2007. Aberrant infection and persistence of varicella-zoster virus in human dorsal root ganglia *in vivo* in the absence of glycoprotein I. *Proc Natl Acad Sci U S A,* 104, 14086-91.

ZERBONI, L., SEN, N., OLIVER, S. L. & ARVIN, A. M. 2014. Molecular mechanisms of varicella zoster virus pathogenesis. *Nat Rev Microbiol,* 12, 197-210.

CHAPTER 6

Epstein-Barr Virus Infections and Vaccine Advances

Risaku Fukumoto*

Scientific Research Department, Armed Forces Radiobiology Research Institute, Uniformed Services University of the Health Sciences, Bethesda, MD, USA

Abstract: Epstein-Barr virus (EBV) infection is associated with a spectrum of fatal diseases including Burkitt's lymphoma (BL) and nasopharyngeal carcinoma (NPC). Higher incidence of BL and NPC seen in developing world demands an effective vaccine to prevent infection and disease onset. EBV primarily infects people during infant or adolescent ages and establishes latent infection. Vast majority of EBV-associated diseases develop later in immune-compromised individuals, suggesting the link between host immunity and viral reactivation. This chapter discusses viral factors that EBV produces whose interactions with host may determine infection, reactivation, immune evasion and disease progression, and further reviews vaccine formulations tested to date. Knowledge in such viral factors is indispensable for vaccine advances.

Keywords: Epstein-Barr virus, Burkitt's lymphoma, viral factors, latent infection, nasopharyngeal carcinoma, Herpesviridae, *Gammaherpesvirinae*, B cells, infectious mononucleosis, EBNA, LMP, lytic phase, latency, EBER, BCRF1, BHRF1, BZLF1, gp350, gp220, virus

INTRODUCTION

It has been half century since discovery of Epstein-Barr virus (EBV). EBV infection causes Burkitt's lymphoma (BL), nasopharyngeal carcinoma (NPC) and several other lymphoproliferative disorders (Cohen, 2000, Kutok and Wang, 2006). Vaccine to EBV has been a pressing need especially in developing world where EBV-associated diseases are endemic and often lethal (Cohen *et al.*, 2011). EBV belongs to *Herpesviridae*, a large family in Group I (dsDNA) viruses that infects humans and animals and causes a variety of diseases. EBV belongs to *Herpesviridae* subfamily *Gammaherpesvirinae* and genus *Lymphocryptovirus* together with several others such as pongine herpes viruses 1-3 found in non-

---

*Corresponding author Risaku Fukumoto: Scientific Research Department, Armed Forces Radiobiology Research Institute, Uniformed Services University of the Health Sciences, 8901 Wisconsin Avenue, Bethesda, Maryland 20889-5603, USA; E-mail: risaku_fukumoto@hotmail.com

human primates. EBV is also sometimes referred to as human herpesvirus-4 (HHV-4), among eight HHVs that have been identified to infect humans.

In 1958, British surgeon Burkitt found characteristic lymphoma, which is now known as BL, in children in equatorial Africa where malaria is endemic and suggested the presence of infectious agent that caused them (Burkitt, 1969, Burkitt, 1971). EBV viral particles were discovered in 1964 by British virologists Epstein and Barr together with an electron microscopist Achong in cultured cells derived from a BL biopsy (Epstein *et al.*, 1964).

Although many cases of BL contain EBV, this virus was later found ubiquitous and in most cases asymptomatic: it has been estimated that more than 95% of world adult population are infected by EBV (Baumforth *et al.*, 1999, Andersson, 2000). This suggested that the development of diseases is largely dependent on host immunity that may control viral reactivation. This hypothesis is supported by the facts including higher incidence of primary central nervous system lymphoma (PCNSL) developing in AIDS patients (MacMahon *et al.*, 1991) and post-transplant lymphoma (PTL) in immune depleted individuals (Young *et al.*, 1989).

PRIMARY INFECTION

Primary infection by EBV usually occurs in infancy as soon as the maternal antibody to EBV disappears, but it can also appear later in life, typically during adolescent ages (Fleisher *et al.*, 1979). Oropharyngeal epithelial cells are believed to serve as primary EBV infection and replication site (Sixbey *et al.*, 1984). Viruses that are in their lytic phase in these cells can spread through saliva and oral mucosa of the donor individuals as cell-free particles. Other individuals become infected through exchange of saliva or contact of oral mucosa with viral donor. Most commonly, infection occurs in infant's day care settings where mouthing of toys is common, or during adolescence with frequent kissing. Both of these routes potentially involve interactions with random, multiple viral donors and constitute infection peaks during the lifespan. Other routes of infection that have been reported are through close sexual contacts (Crawford *et al.*, 2002) and blood transfusions (Wagner *et al.*, 1994), and virus has been found in semen, breast milk and cervical epithelium (Israele *et al.*, 1991, Junker *et al.*, 1991, Sixbey *et al.*, 1986). Possible perinatal infection has also been reported (Meyohas *et al.*, 1996).

LATENT INFECTION

After primary infection, EBV establishes latent infection in B-lymphocytes – a lifelong reservoir – and through them circulates and colonizes multiple organs including tonsillar (Young and Rickinson, 2004), lymph nodes (Shibata *et al.*, 1991), oral mucosa (Ammatuna *et al.*, 1998) and genital tract (Israele *et al.*, 1991). EBV produced in oropharyngeal epithelial cells infects B-lymphocytes through engagement of viral envelope glycoproteins gp350 and gp42 with a cell surface receptor CD21 and a co-receptor, major histocompatibility complex (MHC) class II, that are specifically present on B-lymphocytes (Tanner *et al.*, 1987, Nemerow *et al.*, 1987). Since viral replication in B-lymphocytes depletes gp42 due to MHC II-mediated internalization, the virions produced by B-lymphocytes are free of gp42 and are selectively infecting epithelial cells rather than other B-lymphocytes (Borza and Hutt-Fletcher, 2002). This is considered to be a mechanism in which EBV switches its tropism between these two populations. EBV can also infect *in vivo* T-cells and natural killer (NK) cells in few cases by unknown mechanisms (Jones *et al.*, 1988, Harabuchi *et al.*, 1990), and do cause lymphomas in these subsets (see 'diseases' section).

CLINICAL MANIFESTATIONS

Primary infection with EBV causes infectious mononucleosis (IM), while latent infection is associated with many diseases that include Burkitt's lymphoma (BL), nasopharyngeal carcinoma (NPC), post-transplant lymphoma (PTL), primary central nervous system lymphoma (PCNSL), Hodgkin's lymphoma (HL), and oral hairy leukoplakia (OHL). The fact that many lymphoproliferative disorders occur during primary infection or upon immune suppression underline importance of host adaptive immunity in controlling EBV infection. There is currently no specific treatment to any EBV-associated diseases except prevention and treatment of secondary infections by antimicrobials or symptomatic therapy (Gershburg and Pagano, 2005).

Infectious mononucleosis (IM): IM, also known as Pfeiffer's Disease or colloquially as "kissing disease", occurs in 30-40% of adolescents (>15 y.o.) during the primary infection with EBV (Fleisher *et al.*, 1979). Typical symptoms include fever, fatigue, pharyngitis and cervical lymphadenopathy and occasionally more serious complications may follow, such as fulminant hepatic failure, splenic rupture, and hematologic disorders (Sokal *et al.*, 2007). The infections that occur in adolescence and adulthood may exhibit several symptoms described above whereas in infant none or very modest symptoms are normally observed and if any, disappear in a few days with no treatment (Fleisher *et al.*, 1979). Patients with IM shed in saliva infectious viruses that are in their lytic phase (Young and Rickinson, 2004). Pronounced lymphocytosis with atypical T-lymphocytes has been found in this condition (Sheldon *et al.*, 1973, Hislop *et al.*, 2002).

Burkitt's lymphoma (BL): BL is a malignancy of B-lymphocytes. According to World Health Organization (WHO) classification, there are three clinical subtypes of BL: *endemic, sporadic* and *immunodeficiency-associated* (Young and Rickinson, 2004, Ferry, 2006). The onset of all BL is believed to involve impaired host immunity at different levels, where reactivation of viruses can occur. The incidence of *endemic* BL is high in central Africa and New Guinea where chronic malaria overlaps, and EBV was firstly discovered in a BL biopsy isolated in this region. EBV is found in all *endemic* BL biopsies. It is suggested that malaria renders the infected individual to become more susceptible to EBV reactivation (Epstein, 1984, Chene *et al.*, 2007). *Sporadic* BL occurs at much lower rates in developed countries and <30% cases contain EBV (Burmeister *et al.*, 2005). There is also *intermediate* BL reported (*i.e.* both geographic prevalence and EBV positivity are between *endemic* and *sporadic*). This subtype is found in North Africa and Brazil, and EBV is present in <85% of the cases (Young and Rickinson, 2004). The relation between *sporadic/intermediate* BL and host immunity is currently undefined. In addition, *Immunodeficiency-associated* BL is common in HIV[+] individuals and fewer in patients with immunosuppressive treatment after organ transplant. It has been reported that BL is 1,000 times more common in HIV[+] individuals than in general population, and that 30-40% of non-Hodgkin's lymphoma in HIV[+] patients are BL (Ferry, 2006). All BL subtypes are highly associated with chromosomal translocations, most frequently t(8;14), that dysregulate *c*-myc expression, suggesting its role in BL development (Rabbitts *et al.*, 1984b, Polack *et al.*, 1996, Li *et al.*, 2003, Rabbitts *et al.*, 1984a, Jacobson and LaCasce, 2014).

Nasopharyngeal carcinoma (NPC): NPC is a malignancy of nasopharyngeal epithelial cells. EBV DNA is detectable in more than 90% of the NPC cases (Fahraeus *et al.*, 1988). Thus, EBV infection is considered to be strongly associated with NPC. There seems to be genetic and dietary factors that favor development of NPC. Most frequent occurrence of NPC is located in a part of China and expatriate Chinese, mostly Cantonese males, regardless of the place where they live (Yu and Yuan, 2002). Particular diet such as salty fish that is common in this population may be a co-factor that favors disease progression (Yu *et al.*, 1986). Clinically, serological screenings using peripheral blood to determine IgA titer against EBV (Henle and Henle, 1976, Ho *et al.*, 1976, Zeng, 1985) as well as viral DNA load (Chan *et al.*, 2002) are two important parameters because these parameters correlate well with NPC progression statuses and prognosis, depending upon the timing of examinations before or after the treatments. Thus, the lack of immunity to EBV may play a role in viral reactivation and NPC development, too.

Post-transplant lymphoma (PTL): T-cell immune function plays a major role in keeping EBV-mediated lymphoma development under control, and this is best exemplified by observing patients receiving immunosuppressant after organ transplant (Young *et al.*, 1989, Zimmermann and Trappe, 2013) and also patients with AIDS (see also 'PCNSL'). Both of these groups are at a greater risk of developing lymphoma presumably by similar mechanism (Young and Rickinson, 2004). During PTL progression, early lesions appear within a year of transplant. They consist of monoclonal and polyclonal expansions of EBV$^+$ B-cells and these cells later undergo malignant transformation (Young and Rickinson, 2004). EBVs in lymphoma cells are in their latency program III (see 'viral latency and diseases' section).

Primary central nervous system lymphoma (PCNSL): PCNSL is a lymphoma that develops in the lymph tissues of the brain and spinal cord (Miller *et al.*, 1994). It consists mostly of malignant EBV$^+$ B-cells and of EBV$^+$ NK/T-cells in very few cases (Kaluza *et al.*, 2006). Similarly to PTL, EBV infection in general causes PCNSL only in immunodeficient individuals including AIDS and those who receive immunosuppressant (MacMahon *et al.*, 1991, Vernino *et al.*, 2005). This is one of the best examples where loss of CD4$^+$ T-cell-mediated immunity could directly promote EBV-induced lymphoma progression. EBV transcripts EBER1 were detected in PCNSL biopsies of AIDS patients (MacMahon *et al.*, 1991).

T-cell and NK-cell lymphomas: EBV can infect CD4$^+$ T-cells, CD8$^+$ T-cells and also NK cells in very few incidences and can transform them into EBV$^+$ T-cell and NK-cell lymphomas (Jones *et al.*, 1988, Harabuchi *et al.*, 1990). These unique lymphomas are most commonly seen in the Southeast Asian population, suggesting genetic disposition that favors disease development. Mechanisms for this remain unknown. Viral expression patterns in these lymphomas are those of latent phases II or I (see 'viral latency and diseases' section), with variable expression of LMP1 in only part of the population (Young and Rickinson, 2004).

Oral hairy leukoplakia (OHL): It typically appears as white mucosal plaque on both sides of tongue of HIV-infected patients. This condition is caused by uncontrolled, opportunistic replication of EBV as a result of severe immune deficiency in AIDS patients (Greenspan *et al.*, 1985). Hairy leukoplakia can also be seen in patients under immune suppressive drugs and steroid treatments (Piperi *et al.*, 2010). This condition does not progress to malignancy.

Gastric carcinoma: It has been reported that 16% of gastric adenocarcinoma contained EBV genes detected by polymerase chain reaction (Shibata and Weiss,

1992). Other groups in different regions in the world support this initial report with lower frequency in positivity (6.9% and 5.6%), in which the EBERs (see 'viral gene products' section) were detected by *in situ* hybridization (Tokunaga *et al.*, 1993, Lee *et al.*, 2004). Thus, the different positivity between analyses may be due to biopsies' ethnic background and/or detection methods used. In all reports, incidence of EBV$^+$ gastric carcinoma is significantly higher in males than in females. EBV$^+$ gastric carcinoma has different phenotype compared to EBV$^-$ ones (Lee *et al.*, 2004). The role of EBV in the development of this carcinoma has not been elucidated. Because EBV is not found in pre-malignant lesions, this could mean that the infection occurs at later stages in carcinoma development (Tokunaga *et al.*, 1993).

Hodgkin's lymphoma (HL): HL occurs at the highest rate in male children under the age 10, and in developed countries EBV is found in 40% to 50% of all HL cases (Weiss *et al.*, 1989, Weiss, 2000, Jarrett *et al.*, 1991). Although the exact role in lymphomagenesis remains unclear, EBV in HL is linked to p21 suppression and worse prognosis (Liu *et al.*, 2010).

VIRAL STRUCTURE AND STRAINS

The viral virion has approximately 100-110 nm in diameter and consists of icosahedral capsid coated by envelope with viral glycoprotein spikes on its surface. The viral envelope's glycoproteins gp350 and gp220 are major membrane antigens (MA), and dominate other much less expressed MAs including gp85, gp25, and also alternatively processed gp42/38. (Li *et al.*, 1995). Gp350 is also referred to gp340 in many reports because of size variations appearing in the analytic gel. Each viral capside particle contains a single copy of about 172 kilobase (kb) of linearized double-stranded DNA genome (Baumforth *et al.*, 1999). The genome size is somewhat variable because of different numbers of 0.5 kb repetitive sequence at both ends named terminal repeats (TRs), as well as the 3.1 kb repeatable spacer sequence, that divides the genome into long and short arms (Baumforth *et al.*, 1999). Upon infection, TRs join to create a circular viral genome (episomal form). One infected cell harbors normally multiple copies of episomal forms. During latent infection, episomal forms are replicated without linearization (Baumforth *et al.*, 1999). Chromosomal integration of linearized EBV genome has also been reported both with or without presence of the episomal forms (Henderson *et al.*, 1983). However, this form is uncommon and integration sites are non-specific. There are two major strains of EBV, EBV-1 and EBV-2 that were formerly also referred to as EBV-A and EBV-B (Young *et al.*, 1987). These two differ in the viral genome regions that encode for the EBNA2

(Adldinger *et al.*, 1985), EBNA3 (Rowe *et al.*, 1989, Sample *et al.*, 1990) and EBER2 (Arrand *et al.*, 1989) (see 'viral gene products' and 'viral factors and persistence' sections). EBV-1 is more effective in causing B-cell outgrowth and proliferation compared to EBV-2 *in vitro* (Baumforth *et al.*, 1999). This is mainly a result of differences in the gene encoding EBNA2 (Rickinson *et al.*, 1987). Despite these *in vitro* transformation abilities, about half of African BL tumors carry EBV-2 (Young *et al.*, 1987).

VIRAL GENE PRODUCTS

EBV genome encodes about 80 genes whose expressions result in production of viral proteins and transcripts (Baumforth *et al.*, 1999). Best known EBV viral proteins include the EBV-encoded nuclear antigens (EBNAs) and the EBV-encoded latent membrane proteins (LMPs) whose expression patterns define viral latency programs (see 'viral latency and diseases' section). Two promoters Wp and Cp, in combination with alternative splicing, operate expression of all EBNAs, resulting in EBNA1, EBNA2, EBNA3A, EBNA3B, EBNA3C and EBNA leader protein (EBNA-LP) (Elliott *et al.*, 2004). There is another promoter Qp that operates only EBNA1 (Nonkwelo *et al.*, 1996). EBNAs that belong to one EBV variant are different from those of other variants (Young *et al.*, 1987). LMPs include LMP1, LMP2A and LMP2B and, unlike EBNAs, are under the regulation of three distinct promoters. Additionally, the virus expresses non-polyadenylated transcripts that include Epstein-Barr early RNAs (EBERs) whose functions may be of significance during EBV-induced pathogenesis.

Viral Latency and Clinical Manifestations

In its lytic phase EBV expresses many viral genes and produces virions outside the host cell, but once it establishes latent infection it becomes dormant (latent phase). Lytic and latent phases in relation to viral gene expressions are described below.

Lytic phase: EBV in this phase is seen transiently after primary infection or upon opportunistic reactivation of latent viruses. Given the absence of host immunity during primary infection or in immunocompromised conditions, the virus in oral epithelial and B-cells expresses two key viral transactivators BZLF1 and BRLF1 that trigger sequential expression of numerous viral antigens (Baumforth *et al.*, 1999, Tsurumi *et al.*, 2005). Such antigens include early antigen (EA), viral capsid antigen (VCA) and membrane antigen (MA) complexes. They are unique to the lytic phase and are required for viral structural components, therefore for viral replication. Viral replication can occur with or without host cell lysis (Sista *et al.*, 1995). Produced viral virions in oral mucosa are infectious and can be transmitted to other individuals through secretion in saliva, or can cause repeated infection of B-lymphocytes within the host, which results in the next latent incubation period.

Latent phase: After primary infection or reactivation, EBV transits to latent phase in B-lymphocytes. Latent phase is believed to be important for the virus to persist when host immunity becomes dominant. The virus limits expression of its own genes in order to avoid recognition by the immune system. There are several distinct gene expression patterns that define latency programs 0-III as below (Rowe *et al.*, 1992, Brooks *et al.*, 1993). Importantly, each latency program is associated with specific malignancies that EBV causes.

Latency 0	EBERs		
Latency I	EBER1, 2	EBNA1	
Latency II	EBER1, 2	EBNA1	LMP1, 2A/2B
Latency III	EBER1, 2	EBNA1, 2, 3A/3B/3C, -LP	LMP1, 2A/2B

Latency 0: It has been proposed that all viral genes are dormant in normal resting infected B-cells (Young and Rickinson, 2004, Hochberg *et al.*, 2004). However, EBERs may be detected (Shaknovich *et al.*, 2006).

Latency I: This latency program was found in cell lines that continuously retained their *in vivo* phenotype classified as "group I BL", which is majority in all BLs (Hochberg *et al.*, 2004, Gregory *et al.*, 1990, Gregory *et al.*, 1991). EBNA1 is the only viral protein that can be detected in this latency.

Latency II: This is the latency program most recently found. It is interesting that latency II is found in non-B-cell malignancies such as NPC (Brooks *et al.*, 1993), HL (Deacon *et al.*, 1993) and T-cell (Yoshiyama *et al.*, 1995, Minarovits *et al.*, 1994) and NK-cell lymphomas (Chiang *et al.*, 1996, Kieff, 1996).

Latency III: This is the first identified latency program, found in "group III BL" and EBV-transformed lymphoblastoid cell lines (LCLs), with expression of all EBNAs and LMPs (Kerr *et al.*, 1992). It has been known that many BL biopsies of group I phenotype when transferred to *in vitro* culture change their phenotypes as well as the viral latency program to III (Kerr *et al.*, 1992, Gregory *et al.*, 1990). Latency III is also common in PTLs of the patients who received immunosuppressant and PCNSL of AIDS patients (Young *et al.*, 1989). Thus, evidence supports the presence of T-cell immunity that controls "latency III-associated" lymphoproliferative diseases (*i.e.* such immunity is absent *in vitro* or deficient in AIDS).

VIRAL FACTORS INFLUENCING PERSISTENCE

EBV succeeds to persist for life by shedding several virulent factors. First, these factors sabotage host immune surveillance and promote viral immune evasion by a mechanism similar to other herpes viruses. For example, both EBV and cytomegalovirus encode interleukin-10 (IL-10) homologue that suppresses host immunity (Slobedman *et al.*, 2009, Hsu *et al.*, 1990, Spencer *et al.*, 2002), and disable MHC class I-restricted antigen presentation achieving immune evasion (Zeidler *et al.*, 1997, Lin *et al.*, 2007). In addition, EBERs induce IL-10 production from host cells (Samanta *et al.*, 2008). Second, EBV protects its host cell from apoptosis by producing BHRF1 during its lytic phase. Third, EBV replicates *in vivo* in a tight self-limiting manner utilizing its positive and negative regulators, BZLF1, EBNA1, LMP2A and LMP2B. Collectively, these ongoing studies suggest that viral factors together promote viral persistency while EBV ultimately achieves its spread in host.

BCRF1: BCRF1 shares homology with IL-10, and often referred to as *v*-IL-10 (de Waal Malefyt *et al.*, 1991). BCRF1 is secreted during the lytic cycle, and, similar to IL-10, inhibits interferon-gamma (IFN-γ) production by Th1 subsets. (Hsu *et al.*, 1990). Because IFN-γ is required for macrophage activation that kills EBV-infected cells, loss of IFN-γ results in viral immune evasion. In the host cell, BCRF1 also suppresses transcription of transporter associated with antigen presentation 1 (TAP1) gene (Zeidler *et al.*, 1997), which is a component of the machinery responsible for the transportation of foreign antigen peptides to the endothelial reticulum (ER) including those of EBV after their proteolyses. BCRF1 by this impairs antigen loading on MHC I in the ER, diminishes host's ability to present viral antigen on the cell surface, and protects EBV-infected cells from cytotoxic T-lymphocytes (CTLs). Consequently, this dramatically reduces MHC I on cell surface because unloaded MHC I molecules have much shorter half-life and become targeted for degradation (Zeidler *et al.*, 1997).

BHRF1: BHRF1 shares homology with an anti-apoptotic oncoprotein Bcl-2, and similar to Bcl-2 protects cells with EBV in lytic phase from undergoing apoptosis (Henderson *et al.*, 1993). Besides this effect, *in vitro* study comparing epithelial cells stably expressing BHRF1 and Bcl-2 indicates that BHRF1 confers higher rate of proliferation than Bcl-2 does, evidenced by enhanced cell cycle transition (Dawson *et al.*, 1998). These multiple effects may promote survival and therefore persistence of EBV-infected cells, and meantime make them more susceptible to genetic instability and oncogenic transformation in longer period of time.

BZLF1: Stimulation of B-lymphocytes with latent EBV infection by ligation of B-cell antigen receptor (BCR) (Tovey *et al.*, 1978, Takada, 1984), transforming growth factor-β (TGF-β) (di Renzo *et al.*, 1994), phorbol 12-*O*-tetradecanoylphorbol-13-acetate (TPA) (zur Hausen *et al.*, 1978) and sodium butyrate (SB) (Luka *et al.*, 1979) can result in viral reactivation. A key mechanism that switches the virus from latent phase to lytic cycle is controlled by a viral transcription factor BZLF1 (Flemington and Speck, 1990, Lieberman and Berk, 1990). BZLF1 has significant homology with cellular transcription factors Jun-Fos family that binds AP-1 sites on DNA. Upon expression, BZLF1 through its *C*-terminus dimerizes and binds to Z-responsive elements located upstream of target viral genes and by doing so induces "an ordered cascade of early and late lytic genes" (Sinclair and Farrell, 1992, Tsurumi *et al.*, 2005) (See also 'viral latency and diseases' section). BZLF1 can also induce host genes through binding to AP-1 sites (Farrell *et al.*, 1989). *N*-terminal end of BZLF1 associates with a general transcription factor TFIID, a component of TATA box binding protein (TBP) and, is essential to regulate its gene transcriptions (Lieberman and Berk, 1991). It has been demonstrated that BZLF1 can either activate or repress viral promoters by changing host-binding partners. For example, the association with 53BP1, a DNA damage protein activated during EBV replication, can induce viral reactivation (Bailey *et al.*, 2009) while the association with retinoic acid receptor (RAR) can repress it (Sista *et al.*, 1995). Thus, it is important to identify viral and cellular factors that bind BZLF1 and ultimately control lytic activation.

EBNA1: EBNA1 is the only viral antigen that is expressed during the lytic phase and all latencies, except latency 0. It activates viral episomal DNA replication (Yates *et al.*, 1985, Kirchmaier and Sugden, 1995, Rawlins *et al.*, 1985). Molecular weight of the EBNA1 considerably varies between 69-94 kDa depending on viral variant, but no link between size and function is reported (Allday and MacGillivray, 1985, MacGillivray *et al.*, 1988). EBNA1 functions as a homodimer and a *C*-terminal domain is required for both dimerization and DNA binding (Ambinder *et al.*, 1991, Chen *et al.*, 1994). Dimerized EBNA1 binds specifically to two binding sites within origin of DNA replication (*ori*P), a 1.7kbp of *cis*-acting EBV genome region (Kirchmaier and Sugden, 1995, Jones *et al.*, 1989). EBNA1 also binds to the other sequence existing on viral genome DNA (BamHI-Q locus) and a viral RNA transcript EBER1 (Snudden *et al.*, 1994). Thus, EBER1 has potential ability to control EBNA1 and viral replication, and it is abundantly present in BL and other EBV-immortalized primary human B-cells (see 'EBERs'). In addition, EBNA1 has also been implicated in its oncogenic potential because of its ability to induce B-cell lymphomas in transgenic mice (Wilson *et al.*, 1996) and BL survival (Kennedy *et al.*, 2003).

LMP2A and LMP2B: Other important viral proteins that play critical roles in viral regulation are LMP2A and LMP2B, with their ability to alter signal transduction through B-cell receptor (BCR). Because of this, they control viral BZLF1 expression, which triggers lytic transition (see 'BZLF1'). LMP2A and LMP2B share 12 common transmembrane domains (TMs) with difference at their *N*-termini, resulting from alternatively spliced 1st exons in transcripts (Scholle *et al.*, 2000). LMP2A has 119-amino-acid cytoplasmic tail at its *N*-terminus, which carries 8 tyrosine residues that become phosphorylated upon BCR stimulation and two of them serve as immunoreceptor tyrosine-based activation motif (ITAM). Upon phosphorylation, ITAM recruits Src member Lyn and also Syk kinases and inhibits signaling from BCR and reduces BZLF1-mediated viral expression (Caldwell *et al.*, 1998). LMP2B, by lacking *N*-terminus tail, weakens the activity of LMP2A upon expression and restores viral activation. Interestingly, these "positive" and "negative" viral regulators are under the regulation of two different promoters. This may confer the virus to control opportunistic activation mediated by different transcription factors. Another important outcome of LMP2A expression was observed in transgenic mice whose B-cells bear this protein. LMP2A abrogated the normal B-cell development in these mice, which resulted in immunoglobulin negative cells to outgrowth and colonize peripheral organs (Caldwell *et al.*, 1998). This indicates LMP2A blocks BCR signal whilst promotes proliferation and survival. Importantly, TMs common in both LMP2A and LMP2B are required for their co-localization to the cholesterol-sphingolipid rich component termed "lipid-rafts" on the cell surface, to which BCR and signaling molecules are also recruited (Dykstra *et al.*, 2001). No efficient viral regulation can be achieved without this membrane localization.

EBERs: EBER1 (166 bases) and EBER2 (172 bases) are non-polyadenylated and therefore non-coding RNAs that are expressed in all forms of latency and in all

EBV-associated diseases known (Minarovits *et al.*, 1992, Young and Rickinson, 2004). EBERs are widely used for detection of cells latently infected with EBV because of their abundance with up to 10^7 copies / cell (Howe and Steitz, 1986, Baumforth *et al.*, 1999). EBERs are not essential for the EBV-mediated transformation of primary B-cells (Swaminathan *et al.*, 1991). However, when expressed in BL lines they have the abilities to induce IL-10 production, suggesting its role in immune suppression that promotes BL survival *in vivo* (Samanta *et al.*, 2008). Others also reported EBERs bind to and inhibit double-stranded-RNA-activated protein kinase (PKR), which mediates type I interferon-driven apoptosis in BL (Nanbo *et al.*, 2002). EBERs do so by forming stable RNA-protein complex with the autoantigen La (Lerner *et al.*, 1981) and ribosomal protein L22 (Toczyski and Steitz, 1991, Toczyski *et al.*, 1994). EBERs exhibit striking similarities with adenoviral VA RNAs and can functionally substitute them for similar PKR inhibition, which occurs during natural adenoviral infection (Rosa *et al.*, 1981, Bhat and Thimmappaya, 1983). Thus, EBERs' role in immune suppression might be more significant than it has been thought. It is interesting that EBNA1 may induce EBERs expression (Owen *et al.*, 2010). Of note, EBERs and EBNA1 is the only viral gene products found in latency I and may regulate each other (see 'EBNA1').

ROLE OF VIRAL PROTEINS IN PROGRESSION TO BURKITT'S LYMPHOMA

Viral factors regulate both viral and host genes during latent infection that leads to indefinite proliferation and acquisition of neoplastic transformation, most notably BL. Accumulated studies evidence that several viral antigens induce gene expressions characteristic to BL and others are required for transformation processes. It has been observed *in vitro* that LMP1, EBNA2 and EBNA3C when individually expressed induce phenotypic changes that partially resemble BL, and might be important for immortalization (Wang *et al.*, 1990a). More detailed analyses of *in vitro* infection studies using several mutant viruses that lack each one of those proteins revealed that LMP1 and EBNA2 are indispensable in transforming the infected cells (Young and Rickinson, 2004, Kieff and Rickinson, 2001). Strong activation of *c*-myc is observed in all BL cases, suggesting BL on its development acquires host changes too.

LMP1: It is believed that LMP1 plays major roles in transformation of infected B-cells. When individually expressed, it can transform rodent cell lines *in vitro* and induce gene expression similar to BL cells (Wang *et al.*, 1985). Like LMP2A and LMP2B, LMP1 is located on lipid-rafts on cell surface by its 6 transmembrane domains (Kaykas *et al.*, 2001). LMP1 lacks extracellular domain but has long cytoplasmic domain at its *C*-terminus that functionally resembles CD40, a member of tumor necrosis factor receptor (TNFR) superfamily, and transduces signals in the absence of extracellular ligand. These signals include Nuclear Factor κB (NF-κB), the *c*-Jun *N*-terminal kinase (JNK-AP-1), the mitogen activated protein kinase (MAPK), and the Janus kinase-signal transducers and activators of transcription (JAK-STAT) pathways. By activating them, LMP1 induces cell surface molecules CD23, CD39, CD40 and CD44, increases cell surface adhesion molecules CD11a (LFA1), CD54 (ICAM1) and CD58 (LFA3), decreases cell surface expression of CD10 (Baumforth *et al.*, 1999), increases cytokines interleukin-6 and interleukin-10 (Eliopoulos *et al.*, 1997, Nakagomi *et al.*, 1994) and increases anti-apoptotic molecules Bcl-2, MCL-1 and A20 (Gregory *et al.*, 1991, Henderson *et al.*, 1991, Laherty *et al.*, 1992). All these effects together are believed to confer the EBV-infected cells the abilities to clump, proliferate, survive under host immunity and protect them from apoptosis during transformation processes.

EBNA2: EBNA2 plays central roles in viral transformation. It interacts with cellular sequence-specific DNA binding protein Jκ-recombination-binding protein (RBP-Jκ) on its repression domain and disables repression activity (Grossman *et al.*, 1994, Hsieh and Hayward, 1995). EBNA2 by this transactivates LMP1 (Wang *et al.*, 1990b), LMP2A and LMP2B (Zimber-Strobl *et al.*, 1991), and induces cell transformation and control viral replication. EBNA2 also triggers transcription of cellular genes including EBV receptor CD21 (Cordier *et al.*, 1990), and B-cell activation antigen CD23 by the same mechanism (Wang *et al.*, 1987, Wang *et al.*, 1991). During acute infection stage, EBNA2 switches the promoters for EBNAs from Wp to Cp (Woisetschlaeger *et al.*, 1991).

EBNA3A, EBNA3B and EBNA3C: Studies using recombinant EBVs, in that each one of EBNA3 protein is deficient, demonstrated that EBNA3A and EBNA3C but not EBNA3B are critical in transformation of primary B-lymphocytes *in vitro* (Tomkinson and Kieff, 1992, Tomkinson *et al.*, 1993). EBNA3C can transactivate several EBNA2-regulated genes, indicating partial functional overlap (Wang *et al.*, 1990a). However, it also blocks EBNA2-mediated transcription of LMP1 and LMP2 (Le Roux *et al.*, 1994). EBNA3C binds to RBP-Jk so it can compete with EBNA2 upon its expression (Robertson *et al.*, 1995) (see 'EBNA2'). This may be of importance for viral regulation during BL development.

EBNA-LP: EBVs that carry mutant EBNA-LP are partially defective in transforming primary B-cells after infection (Mannick *et al.*, 1991). It is believed that EBNA-LP and EBNA2 are the first antigens to be produced after acute EBV infection to B-cells and together activate cyclin D2 dependent progression of quiescent host into G1 phase (Sinclair *et al.*, 1994).

VACCINE ADVANCES

Majority of EBV-associated diseases are lethal. Thus, development of vaccine has been a pressing need since the discovery of virus. Presently, several key trials indicate prevention of lymphoma in animal models and IM in humans, while none of the proposed vaccines has been licensed (Cohen, 2015). These progresses together with basic research that define molecular details in viral latencies would be needed for advanced vaccines including the ones intended for therapeutic uses.

EBV Antigens in Natural Infection

Natural EBV infection induces humoral and cellular immune response in humans. First, it has been reported that acute EBV infection causes intense antibody response mainly toward viral lytic antigens including MA, EA and VCA (Henle *et al.*, 1970). Second, pronounced lymphocytosis occurs during acute IM, in which atypical cells are T-lymphocytes react to EBV^+ B-cells (Sheldon *et al.*, 1973). Several groups reported CTL reactivity directed to EBNA3A, EBNA3B and EBNA3C (Lee *et al.*, 1997). These antigens are present during lytic phase and latency III, but not in I and II (see 'viral latency and diseases' section). CTLs also recognize other latent antigens including EBNA2 and EBNA-LP (expressed upon acute infection), and also LMP1 and LMP2 (expressed in latency II and barely in latency I) but seem to require more MHC restriction (Baumforth *et al.*, 1999).

In general, latency control by HHVs is a major challenge to vaccine development. EBV persists most of the incubation period in its latency I or 0, in which only EBNA1 and EBERs are expressed, if any (Hochberg *et al.*, 2004). Unfortunately, EBNA1, the only antigen that expresses during any latency (therefore in any EBV-associated malignancy known) was found poorly immunogenic probably because it is highly resistant to MHC presentation (Levitskaya *et al.*, 1995). Thus, it is not realistic to develop vaccines that control latency I or 0. Vaccines that control lytic phase, or latencies III or II, seem to have better opportunity. Current vaccine trials predominantly aim to limit viral replication, and to reduce the risk of IM during primary infection and lymphoma development during latent infection.

Another challenge to EBV vaccine may be polymorphisms seen in EBNA2 and 3 antigens. As discussed earlier, EBNA2 and 3s are known targets of CTLs and produced during latency III. Most importantly, large number of reports support that many EBV-associated malignancies give rise to this latency in immunocompromised individuals, so control over latency III is a key in preventing such diseases. However, each viral strain carries unique EBNAs (Young *et al.*, 1987). One report suggests that immunogenic epitopes in EBNA3B are mutated in an EBV strain existing in Southeast Asia and these mutant epitopes were not recognized by donors who were infected by this mutant strain (de Campos-Lima *et al.*, 1994). Solution to this may require identification of more conserved, immunogenic epitopes or otherwise combination of multiple epitopes.

At present, the use of gp350 may be the most powerful approach, and therefore it is widely used for vaccines against different strains of EBV (see 'subunits'). gp350 is relatively conserved compared to EBNAs (Lees *et al.*, 1993), directly correlates to viral replication (see 'viral latency and diseases' section) and is exposed on the surface of both viral virions and infected cells, allowing easier recognition by immune surveillance. Gp350 can be a major target of EBV-neutralizing antibody (Thorley-Lawson and Poodry, 1982), and CTLs of acute infectious mononucleosis patients exhibit strong *ex vivo* reactivity against gp350 (Khanna *et al.*, 1999). Neutralizing antibodies to gp350 may be beneficial because it may directly inhibit gp350-CD21 engagement required for infection process.

Vaccine Formulations in Animal Models

Several EBV vaccine formulations that have been tested or are planned include: 1) the use of killed- or attenuated-virus, 2) subunit vaccines formulated by pathogen purification or synthetic peptides, or 3) their delivery by viral vectors, to elicit CTL and humoral responses directed to EBV.

Whole virus: Attenuated forms of pathogens have been used for vaccination against several viruses including chickenpox, the only successful vaccine against herpes viruses. This vaccine formulation has advantages for providing immunity the same components of EBV that naturally exist. However, long-term effect of live EBV may be a risk of persistent infections and lymphoma developments. Inactive forms of EBV are poorly immunogenic because they are unable to establish infection and express viral antigens that could be presented on MHC through the cytosolic pathway. Therefore, success in this formulation of EBV vaccine would require safer attenuations, in which viral oncogenicity and immune perturbation are inactivated while retaining infectivity. In conclusion, basic knowledge in viral factors relevant to these functions remains indispensable.

Subunits: Isolation of gp350 from EBV infected cell lines was the initial method in viral antigen preparation for vaccine use (North *et al.*, 1982). It was shown that the gp350 obtained by this method induced immune responses to three different EBV strains (North *et al.*, 1982) and prevented lymphoma development in a cottontop tamarin model (Epstein *et al.*, 1985, Moghaddam *et al.*, 1997), while further human trials required more efficient gp350 production. Production and purification of recombinant gp350, tried in bacteria, yeast and also in mammalian expression systems has replaced direct purification of gp350 from EBV infected cells, but required several improvements because of poor expression or lack/changes in post-translational modifications that reduced immunogenicity (Whang *et al.*, 1987). This is because overexpression of gp350 can be toxic in mammalian cells (Motz *et al.*, 1987) and modifications occur differently in bacteria and yeast (Schultz *et al.*, 1987). Also, expression of naturally existing gp350 ORF on isolated cDNA results in co-expression of gp350 and gp220 on the same frame whose ratio is controlled by alternative splicing (Jackman *et al.*, 1999). One study demonstrated less toxic truncated form of gp350 expressed by bovine-papillomavirus vector being able to induce protective immune response against EBV-induced tumor in cottontop tamarins (Finerty *et al.*, 1992). This membrane-anchor lacking gp350 is efficiently secreted in rodent cell culture supernatants and can be easily purified for vaccine use (Whang *et al.*, 1987, Motz *et al.*, 1987). Further, another group reported the production of gp350 as a single polypeptide chain for vaccine preparation by point mutations that ablate splicing for gp220 production on the less toxic truncated form (Jackman *et al.*, 1999). This gp350 was isolated from cell culture supernatant of stably transfected high-producing clone of Chinese hamster ovary cell (Jackman *et al.*, 1999). Recombinant gp350 subunit vaccines showed optimal efficacy in rabbits (Jackman *et al.*, 1999) and cottontop tamarins (Finerty *et al.*, 1992, Finerty *et al.*, 1994, Morgan *et al.*, 1988a). Thus, studies including above have led gp350 subunit vaccines to advance to human trials (Morgan *et al.*, 1989).

Synthetic peptides: Synthetic peptides often offer easy, fast and cost-effective alternatives in vaccine formulation. However, these peptides lack glycosylation and other modifications that take place on natural EBV antigens *in vivo*, and are thus poorly immunogenic. Therefore, formulations with synthetic peptides would require further improvement in immunogenicity, by selection of highly immunogenic antigens and also by combination with effective immunoadjuvants.

Current immunoadjuvants include immunostimulating complexes (ISCOMS) (Morgan *et al.*, 1988a), a synthetic muramyl dipeptide (MDP) (Morgan *et al.*, 1989) and aluminum salt (Moutschen *et al.*, 2007, Sokal *et al.*, 2007), the only FDA-approved immunoadjuvant in human trials. Presently, these adjuvants have been tested in conjunction with recombinant gp350 but not with synthetic peptides. Animal studies to test the efficacy and safety of peptide-based vaccines may provide hopeful formulations to be further tested in human trials.

Viral vectors: Viral vector vaccines aim at infection of manipulated viruses or bacteria to target cells and make them express EBV antigens of interest. Vaccinia and adenoviral vectors have been used to elicit immune response to gp350, because these viral vectors are used in other vaccines. In 1988 and 1996, recombinant vaccinia virus expressing EBV glycoprotein gp350 was shown to protect cottontop tamarins from EBV-induced malignant lymphomas (Morgan *et al.*, 1988b), and decrease replication of the challenged virus (Mackett *et al.*, 1996). In 1993, replication-defective recombinant adenovirus bearing gp350 was also shown to protect cottontop tamarins from EBV-induced lymphomas that are associated with serum responses to gp350 (Ragot *et al.*, 1993). The use of adenoviral vector may have advantages by its ability to infect tonsils and salivary glands, the proximities of EBV replication. These results may offer the use of viral vectors for gp350 delivery in human.

CLINICAL TRIALS AND THEIR SIGNIFICANCE

Several promising results in animal models reported by different groups revealed that gp350 elicits immune responses, regardless of its delivery methods (*i.e.* uses of subunit and viral vector vaccines), allowing gp350-based vaccine advances to clinical trials. One of the earliest vaccine trials in humans was reported in 1995: in which gp350 was delivered by licensed vaccinia viral vector Tien Tan. Nine infants who were naïve to both EBV and vaccinia were vaccinated and they all developed neutralizing antibody to gp350 successfully. As a result, only three of them became naturally infected by EBV after 16 months, while all 10 unvaccinated infants became infected. Thus, the protection effect of such vaccine from natural EBV infection was reportedly significant (Gu *et al.*, 1995). This study was followed by two more recent sequential reports in 2007 where recombinant gp350 was used (for gp350 see (Jackman *et al.*, 1999)). One report confirmed safety and immunogenicity of recombinant gp350 vaccine in phase I/II settings, tested in a total of 81 healthy, naïve to EBV, volunteers aged 18-37 years (Moutschen *et al.*, 2007). The other report claimed for the first time, the significant effect of EBV vaccine. In this phase II clinical trial, recombinant gp350 vaccine ability to prevent IM was tested in 181 healthy EBV-seronegative adolescent volunteers. The research team reported the prevention of disease with significant efficacy and further seroconversion to gp350 in 98.8% subjects that lasted more than 18 months in length (Sokal *et al.*, 2007). A series of clinical studies indicate that the current gp350-based vaccines are well tolerated, and induce sustained immune response to EBV controlling IM for 18-months. However, the effects of those vaccinations on lifelong viral infection and EBV disease incidence remain to be investigated (Cohen *et al.*, 2013). Answer to this may indicate if alternate vaccine approaches are needed to target other viral components. Current promising results may provide possibility where undesired EBV infection and IM can be avoided, and hopefully further advance to prevention and treatment of more serious diseases, BL and NPC.

CONFLICT OF INTEREST

The author confirms that this chapter contents have no conflict of interest.

ACKNOWLEDGEMENTS

Declared None.

REFERENCES

ADLDINGER, H. K., DELIUS, H., FREESE, U. K., CLARKE, J. & BORNKAMM, G. W. 1985. A putative transforming gene of Jijoye virus differs from that of Epstein-Barr virus prototypes. *Virology,* 141, 221-34.

ALLDAY, M. J. & MACGILLIVRAY, A. J. 1985. Epstein-Barr virus nuclear antigen (EBNA): size polymorphism of EBNA 1. *J Gen Virol,* 66 (Pt 7), 1595-600.

AMBINDER, R. F., MULLEN, M. A., CHANG, Y. N., HAYWARD, G. S. & HAYWARD, S. D. 1991. Functional domains of Epstein-Barr virus nuclear antigen EBNA-1. *J Virol,* 65, 1466-78.

AMMATUNA, P., CAPONE, F., GIAMBELLUCA, D., PIZZO, I., D'ALIA, G. & MARGIOTTA, V. 1998. Detection of Epstein-Barr virus (EBV) DNA and antigens in oral mucosa of renal transplant patients without clinical evidence of oral hairy leukoplakia (OHL). *J Oral Pathol Med,* 27, 420-7.

ANDERSSON, J. 2000. An Overview of Epstein-Barr Virus: from Discovery to Future Directions for Treatment and Prevention. *Herpes,* 7, 76-82.

ARRAND, J. R., YOUNG, L. S. & TUGWOOD, J. D. 1989. Two families of sequences in the small RNA-encoding region of Epstein-Barr virus (EBV) correlate with EBV types A and B. *J Virol,* 63, 983-6.

BAILEY, S. G., VERRALL, E., SCHELCHER, C., RHIE, A., DOHERTY, A. J. & SINCLAIR, A. J. 2009. Functional interaction between Epstein-Barr virus replication protein Zta and host DNA damage response protein 53BP1. *J Virol,* 83, 11116-22.

BAUMFORTH, K. R., YOUNG, L. S., FLAVELL, K. J., CONSTANDINOU, C. & MURRAY, P. G. 1999. The Epstein-Barr virus and its association with human cancers. *Mol Pathol,* 52, 307-22.

BHAT, R. A. & THIMMAPPAYA, B. 1983. Two small RNAs encoded by Epstein-Barr virus can functionally substitute for the virus-associated RNAs in the lytic growth of adenovirus 5. *Proc Natl Acad Sci U S A,* 80, 4789-93.

BORZA, C. M. & HUTT-FLETCHER, L. M. 2002. Alternate replication in B cells and epithelial cells switches tropism of Epstein-Barr virus. *Nat Med,* 8, 594-9.

BROOKS, L. A., LEAR, A. L., YOUNG, L. S. & RICKINSON, A. B. 1993. Transcripts from the Epstein-Barr virus BamHI A fragment are detectable in all three forms of virus latency. *J Virol,* 67, 3182-90.

BURKITT, D. P. 1969. Etiology of Burkitt's lymphoma--an alternative hypothesis to a vectored virus. *J Natl Cancer Inst,* 42, 19-28.

BURKITT, D. P. 1971. Epidemiology of Burkitt's lymphoma. *Proc R Soc Med,* 64, 909-10.

BURMEISTER, T., SCHWARTZ, S., HORST, H. A., RIEDER, H., GOKBUGET, N., HOELZER, D. & THIEL, E. 2005. Molecular heterogeneity of sporadic adult Burkitt-type leukemia/lymphoma as revealed by PCR and cytogenetics: correlation with morphology, immunology and clinical features. *Leukemia,* 19, 1391-8.

CALDWELL, R. G., WILSON, J. B., ANDERSON, S. J. & LONGNECKER, R. 1998. Epstein-Barr virus LMP2A drives B cell development and survival in the absence of normal B cell receptor signals. *Immunity,* 9, 405-11.

CHAN, A. T., LO, Y. M., ZEE, B., CHAN, L. Y., MA, B. B., LEUNG, S. F., MO, F., LAI, M., HO, S., HUANG, D. P. & JOHNSON, P. J. 2002. Plasma Epstein-Barr virus DNA and residual disease after radiotherapy for undifferentiated nasopharyngeal carcinoma. *J Natl Cancer Inst,* 94, 1614-9.

CHEN, M. R., ZONG, J. & HAYWARD, S. D. 1994. Delineation of a 16 amino acid sequence that forms a core DNA recognition motif in the Epstein-Barr virus EBNA-1 protein. *Virology,* 205, 486-95.

CHENE, A., DONATI, D., GUERREIRO-CACAIS, A. O., LEVITSKY, V., CHEN, Q., FALK, K. I., OREM, J., KIRONDE, F., WAHLGREN, M. & BEJARANO, M. T. 2007. A molecular link between malaria and Epstein-Barr virus reactivation. *PLoS Pathog,* 3, e80.

CHIANG, A. K., TAO, Q., SRIVASTAVA, G. & HO, F. C. 1996. Nasal NK- and T-cell lymphomas share the same type of Epstein-Barr virus latency as nasopharyngeal carcinoma and Hodgkin's disease. *Int J Cancer,* 68, 285-90.

COHEN, J. I. 2000. Epstein-Barr virus infection. *N Engl J Med,* 343, 481-92.

COHEN, J. I. 2015. Epstein-barr virus vaccines. *Clin Transl Immunology,* 4, e32.

COHEN, J. I., FAUCI, A. S., VARMUS, H. & NABEL, G. J. 2011. Epstein-Barr virus: an important vaccine target for cancer prevention. *Sci Transl Med,* 3, 107fs7.

COHEN, J. I., MOCARSKI, E. S., RAAB-TRAUB, N., COREY, L. & NABEL, G. J. 2013. The need and challenges for development of an Epstein-Barr virus vaccine. *Vaccine,* 31 Suppl 2, B194-6.

CORDIER, M., CALENDER, A., BILLAUD, M., ZIMBER, U., ROUSSELET, G., PAVLISH, O., BANCHEREAU, J., TURSZ, T., BORNKAMM, G. & LENOIR, G. M. 1990. Stable transfection of Epstein-Barr virus (EBV) nuclear antigen 2 in lymphoma cells containing the EBV P3HR1 genome induces expression of B-cell activation molecules CD21 and CD23. *J Virol,* 64, 1002-13.

CRAWFORD, D. H., SWERDLOW, A. J., HIGGINS, C., MCAULAY, K., HARRISON, N., WILLIAMS, H., BRITTON, K. & MACSWEEN, K. F. 2002. Sexual history and Epstein-Barr virus infection. *J Infect Dis,* 186, 731-6.

DAWSON, C. W., DAWSON, J., JONES, R., WARD, K. & YOUNG, L. S. 1998. Functional differences between BHRF1, the Epstein-Barr virus-encoded Bcl-2 homologue, and Bcl-2 in human epithelial cells. *J Virol,* 72, 9016-24.

DE CAMPOS-LIMA, P. O., LEVITSKY, V., BROOKS, J., LEE, S. P., HU, L. F., RICKINSON, A. B. & MASUCCI, M. G. 1994. T cell responses and virus evolution: loss of HLA A11-restricted CTL epitopes in Epstein-Barr virus isolates from highly A11-positive populations by selective mutation of anchor residues. *J Exp Med,* 179, 1297-305.

DE WAAL MALEFYT, R., HAANEN, J., SPITS, H., RONCAROLO, M. G., TE VELDE, A., FIGDOR, C., JOHNSON, K., KASTELEIN, R., YSSEL, H. & DE VRIES, J. E. 1991. Interleukin 10 (IL-10) and viral IL-10 strongly reduce antigen-specific human T cell proliferation by diminishing the antigen-presenting capacity of monocytes *via* downregulation of class II major histocompatibility complex expression. *J Exp Med,* 174, 915-24.

DEACON, E. M., PALLESEN, G., NIEDOBITEK, G., CROCKER, J., BROOKS, L., RICKINSON, A. B. & YOUNG, L. S. 1993. Epstein-Barr virus and Hodgkin's disease: transcriptional analysis of virus latency in the malignant cells. *J Exp Med,* 177, 339-49.

DI RENZO, L., ALTIOK, A., KLEIN, G. & KLEIN, E. 1994. Endogenous TGF-beta contributes to the induction of the EBV lytic cycle in two Burkitt lymphoma cell lines. *Int J Cancer,* 57, 914-9.

DYKSTRA, M. L., LONGNECKER, R. & PIERCE, S. K. 2001. Epstein-Barr virus coopts lipid rafts to block the signaling and antigen transport functions of the BCR. *Immunity,* 14, 57-67.

ELIOPOULOS, A. G., STACK, M., DAWSON, C. W., KAYE, K. M., HODGKIN, L., SIHOTA, S., ROWE, M. & YOUNG, L. S. 1997. Epstein-Barr virus-encoded LMP1 and CD40 mediate IL-6 production in epithelial cells *via* an NF-kappaB pathway involving TNF receptor-associated factors. *Oncogene,* 14, 2899-916.

ELLIOTT, J., GOODHEW, E. B., KRUG, L. T., SHAKHNOVSKY, N., YOO, L. & SPECK, S. H. 2004. Variable methylation of the Epstein-Barr virus Wp EBNA gene promoter in B-lymphoblastoid cell lines. *J Virol,* 78, 14062-5.

EPSTEIN, M. A. 1984. Burkitt's lymphoma: clues to the role of malaria. *Nature,* 312, 398.

EPSTEIN, M. A., ACHONG, B. G. & BARR, Y. M. 1964. Virus Particles in Cultured Lymphoblasts from Burkitt's Lymphoma. *Lancet,* 1, 702-3.

EPSTEIN, M. A., MORGAN, A. J., FINERTY, S., RANDLE, B. J. & KIRKWOOD, J. K. 1985. Protection of cottontop tamarins against Epstein-Barr virus-induced malignant lymphoma by a prototype subunit vaccine. *Nature,* 318, 287-9.

FAHRAEUS, R., FU, H. L., ERNBERG, I., FINKE, J., ROWE, M., KLEIN, G., FALK, K., NILSSON, E., YADAV, M., BUSSON, P. & ET AL. 1988. Expression of Epstein-Barr virus-encoded proteins in nasopharyngeal carcinoma. *Int J Cancer,* 42, 329-38.

FARRELL, P. J., ROWE, D. T., ROONEY, C. M. & KOUZARIDES, T. 1989. Epstein-Barr virus BZLF1 trans-activator specifically binds to a consensus AP-1 site and is related to c-fos. *EMBO J,* 8, 127-32.

FERRY, J. A. 2006. Burkitt's lymphoma: clinicopathologic features and differential diagnosis. *Oncologist,* 11, 375-83.

FINERTY, S., MACKETT, M., ARRAND, J. R., WATKINS, P. E., TARLTON, J. & MORGAN, A. J. 1994. Immunization of cottontop tamarins and rabbits with a candidate vaccine against the Epstein-Barr virus based on the major viral envelope glycoprotein gp340 and alum. *Vaccine,* 12, 1180-4.

FINERTY, S., TARLTON, J., MACKETT, M., CONWAY, M., ARRAND, J. R., WATKINS, P. E. & MORGAN, A. J. 1992. Protective immunization against Epstein-Barr virus-induced disease in cottontop tamarins using the virus envelope glycoprotein gp340 produced from a bovine papillomavirus expression vector. *J Gen Virol,* 73 (Pt 2), 449-53.

FLEISHER, G., HENLE, W., HENLE, G., LENNETTE, E. T. & BIGGAR, R. J. 1979. Primary infection with Epstein-Barr virus in infants in the United States: clinical and serologic observations. *J Infect Dis,* 139, 553-8.

FLEMINGTON, E. & SPECK, S. H. 1990. Autoregulation of Epstein-Barr virus putative lytic switch gene BZLF1. *J Virol,* 64, 1227-32.

GERSHBURG, E. & PAGANO, J. S. 2005. Epstein-Barr virus infections: prospects for treatment. *J Antimicrob Chemother,* 56, 277-81.

GREENSPAN, J. S., GREENSPAN, D., LENNETTE, E. T., ABRAMS, D. I., CONANT, M. A., PETERSEN, V. & FREESE, U. K. 1985. Replication of Epstein-Barr virus within the epithelial cells of oral "hairy" leukoplakia, an AIDS-associated lesion. *N Engl J Med,* 313, 1564-71.

GREGORY, C. D., DIVE, C., HENDERSON, S., SMITH, C. A., WILLIAMS, G. T., GORDON, J. & RICKINSON, A. B. 1991. Activation of Epstein-Barr virus latent genes protects human B cells from death by apoptosis. *Nature,* 349, 612-4.

GREGORY, C. D., ROWE, M. & RICKINSON, A. B. 1990. Different Epstein-Barr virus-B cell interactions in phenotypically distinct clones of a Burkitt's lymphoma cell line. *J Gen Virol,* 71 (Pt 7), 1481-95.

GROSSMAN, S. R., JOHANNSEN, E., TONG, X., YALAMANCHILI, R. & KIEFF, E. 1994. The Epstein-Barr virus nuclear antigen 2 transactivator is directed to response elements by the J kappa recombination signal binding protein. *Proc Natl Acad Sci U S A,* 91, 7568-72.

GU, S. Y., HUANG, T. M., RUAN, L., MIAO, Y. H., LU, H., CHU, C. M., MOTZ, M. & WOLF, H. 1995. First EBV vaccine trial in humans using recombinant vaccinia virus expressing the major membrane antigen. *Dev Biol Stand,* 84, 171-7.

HARABUCHI, Y., YAMANAKA, N., KATAURA, A., IMAI, S., KINOSHITA, T., MIZUNO, F. & OSATO, T. 1990. Epstein-Barr virus in nasal T-cell lymphomas in patients with lethal midline granuloma. *Lancet,* 335, 128-30.

HENDERSON, A., RIPLEY, S., HELLER, M. & KIEFF, E. 1983. Chromosome site for Epstein-Barr virus DNA in a Burkitt tumor cell line and in lymphocytes growth-transformed *in vitro*. *Proc Natl Acad Sci U S A,* 80, 1987-91.

HENDERSON, S., HUEN, D., ROWE, M., DAWSON, C., JOHNSON, G. & RICKINSON, A. 1993. Epstein-Barr virus-coded BHRF1 protein, a viral homologue of Bcl-2, protects human B cells from programmed cell death. *Proc Natl Acad Sci U S A,* 90, 8479-83.

HENDERSON, S., ROWE, M., GREGORY, C., CROOM-CARTER, D., WANG, F., LONGNECKER, R., KIEFF, E. & RICKINSON, A. 1991. Induction of bcl-2 expression by Epstein-Barr virus latent membrane protein 1 protects infected B cells from programmed cell death. *Cell,* 65, 1107-15.

HENLE, G. & HENLE, W. 1976. Epstein-Barr virus-specific IgA serum antibodies as an outstanding feature of nasopharyngeal carcinoma. *Int J Cancer,* 17, 1-7.

HENLE, W., HENLE, G., ZAJAC, B. A., PEARSON, G., WAUBKE, R. & SCRIBA, M. 1970. Differential reactivity of human serums with early antigens induced by Epstein-Barr virus. *Science,* 169, 188-90.

HISLOP, A. D., ANNELS, N. E., GUDGEON, N. H., LEESE, A. M. & RICKINSON, A. B. 2002. Epitope-specific evolution of human CD8(+) T cell responses from primary to persistent phases of Epstein-Barr virus infection. *J Exp Med,* 195, 893-905.

HO, H. C., NG, M. H., KWAN, H. C. & CHAU, J. C. 1976. Epstein-Barr-virus-specific IgA and IgG serum antibodies in nasopharyngeal carcinoma. *Br J Cancer,* 34, 655-60.

HOCHBERG, D., MIDDELDORP, J. M., CATALINA, M., SULLIVAN, J. L., LUZURIAGA, K. & THORLEY-LAWSON, D. A. 2004. Demonstration of the Burkitt's lymphoma Epstein-Barr virus phenotype in dividing latently infected memory cells *in vivo*. *Proc Natl Acad Sci U S A,* 101, 239-44.

HOWE, J. G. & STEITZ, J. A. 1986. Localization of Epstein-Barr virus-encoded small RNAs by *in situ* hybridization. *Proc Natl Acad Sci U S A,* 83, 9006-10.

HSIEH, J. J. & HAYWARD, S. D. 1995. Masking of the CBF1/RBPJ kappa transcriptional repression domain by Epstein-Barr virus EBNA2. *Science,* 268, 560-3.

HSU, D. H., DE WAAL MALEFYT, R., FIORENTINO, D. F., DANG, M. N., VIEIRA, P., DE VRIES, J., SPITS, H., MOSMANN, T. R. & MOORE, K. W. 1990. Expression of interleukin-10 activity by Epstein-Barr virus protein BCRF1. *Science,* 250, 830-2.

ISRAELE, V., SHIRLEY, P. & SIXBEY, J. W. 1991. Excretion of the Epstein-Barr virus from the genital tract of men. *J Infect Dis,* 163, 1341-3.

JACKMAN, W. T., MANN, K. A., HOFFMANN, H. J. & SPAETE, R. R. 1999. Expression of Epstein-Barr virus gp350 as a single chain glycoprotein for an EBV subunit vaccine. *Vaccine,* 17, 660-8.

JACOBSON, C. & LACASCE, A. 2014. How I treat Burkitt lymphoma in adults. *Blood,* 124, 2913-20.

JARRETT, R. F., GALLAGHER, A., JONES, D. B., ALEXANDER, F. E., KRAJEWSKI, A. S., KELSEY, A., ADAMS, J., ANGUS, B., GLEDHILL, S., WRIGHT, D. H. & ET AL. 1991. Detection of Epstein-Barr virus genomes in Hodgkin's disease: relation to age. *J Clin Pathol,* 44, 844-8.

JONES, C. H., HAYWARD, S. D. & RAWLINS, D. R. 1989. Interaction of the lymphocyte-derived Epstein-Barr virus nuclear antigen EBNA-1 with its DNA-binding sites. *J Virol,* 63, 101-10.

JONES, J. F., SHURIN, S., ABRAMOWSKY, C., TUBBS, R. R., SCIOTTO, C. G., WAHL, R., SANDS, J., GOTTMAN, D., KATZ, B. Z. & SKLAR, J. 1988. T-cell lymphomas containing Epstein-Barr viral DNA in patients with chronic Epstein-Barr virus infections. *N Engl J Med,* 318, 733-41.

JUNKER, A. K., THOMAS, E. E., RADCLIFFE, A., FORSYTH, R. B., DAVIDSON, A. G. & RYMO, L. 1991. Epstein-Barr virus shedding in breast milk. *Am J Med Sci,* 302, 220-3.

KALUZA, V., RAO, D. S., SAID, J. W. & DE VOS, S. 2006. Primary extranodal nasal-type natural killer/T-cell lymphoma of the brain: a case report. *Hum Pathol,* 37, 769-72.

KAYKAS, A., WORRINGER, K. & SUGDEN, B. 2001. CD40 and LMP-1 both signal from lipid rafts but LMP-1 assembles a distinct, more efficient signaling complex. *EMBO J,* 20, 2641-54.

KENNEDY, G., KOMANO, J. & SUGDEN, B. 2003. Epstein-Barr virus provides a survival factor to Burkitt's lymphomas. *Proc Natl Acad Sci U S A,* 100, 14269-74.

KERR, B. M., LEAR, A. L., ROWE, M., CROOM-CARTER, D., YOUNG, L. S., ROOKES, S. M., GALLIMORE, P. H. & RICKINSON, A. B. 1992. Three transcriptionally distinct forms of Epstein-Barr virus latency in somatic cell hybrids: cell phenotype dependence of virus promoter usage. *Virology,* 187, 189-201.

KHANNA, R., SHERRITT, M. & BURROWS, S. R. 1999. EBV structural antigens, gp350 and gp85, as targets for *ex vivo* virus-specific CTL during acute infectious mononucleosis: potential use of gp350/gp85 CTL epitopes for vaccine design. *J Immunol,* 162, 3063-9.

KIEFF, E. 1996. Epstein-Barr virus and its replication. *Fields Virology,* 2, 2343-96.

KIEFF, E. & RICKINSON, A. B. 2001. *Fields Virology,* 2511-74.

KIRCHMAIER, A. L. & SUGDEN, B. 1995. Plasmid maintenance of derivatives of oriP of Epstein-Barr virus. *J Virol,* 69, 1280-3.

KUTOK, J. L. & WANG, F. 2006. Spectrum of Epstein-Barr virus-associated diseases. *Annu Rev Pathol,* 1, 375-404.

LAHERTY, C. D., HU, H. M., OPIPARI, A. W., WANG, F. & DIXIT, V. M. 1992. The Epstein-Barr virus LMP1 gene product induces A20 zinc finger protein expression by activating nuclear factor kappa B. *J Biol Chem,* 267, 24157-60.

LE ROUX, A., KERDILES, B., WALLS, D., DEDIEU, J. F. & PERRICAUDET, M. 1994. The Epstein-Barr virus determined nuclear antigens EBNA-3A, -3B, and -3C repress EBNA-2-mediated transactivation of the viral terminal protein 1 gene promoter. *Virology,* 205, 596-602.

LEE, H. S., CHANG, M. S., YANG, H. K., LEE, B. L. & KIM, W. H. 2004. Epstein-barr virus-positive gastric carcinoma has a distinct protein expression profile in comparison with epstein-barr virus-negative carcinoma. *Clin Cancer Res,* 10, 1698-705.

LEE, S. P., TIERNEY, R. J., THOMAS, W. A., BROOKS, J. M. & RICKINSON, A. B. 1997. Conserved CTL epitopes within EBV latent membrane protein 2: a potential target for CTL-based tumor therapy. *J Immunol,* 158, 3325-34.

LEES, J. F., ARRAND, J. E., PEPPER, S. D., STEWART, J. P., MACKETT, M. & ARRAND, J. R. 1993. The Epstein-Barr virus candidate vaccine antigen gp340/220 is highly conserved between virus types A and B. *Virology,* 195, 578-86.

LERNER, M. R., ANDREWS, N. C., MILLER, G. & STEITZ, J. A. 1981. Two small RNAs encoded by Epstein-Barr virus and complexed with protein are precipitated by antibodies from patients with systemic lupus erythematosus. *Proc Natl Acad Sci U S A,* 78, 805-9.

LEVITSKAYA, J., CORAM, M., LEVITSKY, V., IMREH, S., STEIGERWALD-MULLEN, P. M., KLEIN, G., KURILLA, M. G. & MASUCCI, M. G. 1995. Inhibition of antigen processing by the internal repeat region of the Epstein-Barr virus nuclear antigen-1. *Nature,* 375, 685-8.

LI, Q., TURK, S. M. & HUTT-FLETCHER, L. M. 1995. The Epstein-Barr virus (EBV) BZLF2 gene product associates with the gH and gL homologs of EBV and carries an epitope critical to infection of B cells but not of epithelial cells. *J Virol,* 69, 3987-94.

LI, Z., VAN CALCAR, S., QU, C., CAVENEE, W. K., ZHANG, M. Q. & REN, B. 2003. A global transcriptional regulatory role for c-Myc in Burkitt's lymphoma cells. *Proc Natl Acad Sci U S A,* 100, 8164-9.

LIEBERMAN, P. M. & BERK, A. J. 1990. *In vitro* transcriptional activation, dimerization, and DNA-binding specificity of the Epstein-Barr virus Zta protein. *J Virol,* 64, 2560-8.

LIEBERMAN, P. M. & BERK, A. J. 1991. The Zta trans-activator protein stabilizes TFIID association with promoter DNA by direct protein-protein interaction. *Genes Dev,* 5, 2441-54.

LIN, A., XU, H. & YAN, W. 2007. Modulation of HLA expression in human cytomegalovirus immune evasion. *Cell Mol Immunol,* 4, 91-8.

LIU, T. Y., WU, S. J., HUANG, M. H., LO, F. Y., TSAI, M. H., TSAI, C. H., HSU, S. M. & LIN, C. W. 2010. EBV-positive Hodgkin lymphoma is associated with suppression of p21cip1/wafl and a worse prognosis. *Mol Cancer,* 9, 32.

LUKA, J., KALLIN, B. & KLEIN, G. 1979. Induction of the Epstein-Barr virus (EBV) cycle in latently infected cells by n-butyrate. *Virology,* 94, 228-31.

MACGILLIVRAY, A. J., ALLDAY, M. J., SAUNDERS, S. E. & SINCLAIR, J. H. 1988. EBNA-1: a virally induced nuclear antigen of primate lymphocytes and its expression in Drosophila cells. *Br J Cancer Suppl,* 9, 93-7.

MACKETT, M., COX, C., PEPPER, S. D., LEES, J. F., NAYLOR, B. A., WEDDERBURN, N. & ARRAND, J. R. 1996. Immunisation of common marmosets with vaccinia virus expressing Epstein-Barr virus (EBV) gp340 and challenge with EBV. *J Med Virol,* 50, 263-71.

MACMAHON, E. M., GLASS, J. D., HAYWARD, S. D., MANN, R. B., BECKER, P. S., CHARACHE, P., MCARTHUR, J. C. & AMBINDER, R. F. 1991. Epstein-Barr virus in AIDS-related primary central nervous system lymphoma. *Lancet,* 338, 969-73.

MANNICK, J. B., COHEN, J. I., BIRKENBACH, M., MARCHINI, A. & KIEFF, E. 1991. The Epstein-Barr virus nuclear protein encoded by the leader of the EBNA RNAs is important in B-lymphocyte transformation. *J Virol,* 65, 6826-37.

MEYOHAS, M. C., MARECHAL, V., DESIRE, N., BOUILLIE, J., FROTTIER, J. & NICOLAS, J. C. 1996. Study of mother-to-child Epstein-Barr virus transmission by means of nested PCRs. *J Virol,* 70, 6816-9.

MILLER, D. C., HOCHBERG, F. H., HARRIS, N. L., GRUBER, M. L., LOUIS, D. N. & COHEN, H. 1994. Pathology with clinical correlations of primary central nervous system non-Hodgkin's lymphoma. The Massachusetts General Hospital experience 1958-1989. *Cancer,* 74, 1383-97.

MINAROVITS, J., HU, L. F., IMAI, S., HARABUCHI, Y., KATAURA, A., MINAROVITS-KORMUTA, S., OSATO, T. & KLEIN, G. 1994. Clonality, expression and methylation patterns of the Epstein-Barr virus genomes in lethal midline granulomas classified as peripheral angiocentric T cell lymphomas. *J Gen Virol,* 75 (Pt 1), 77-84.

MINAROVITS, J., HU, L. F., MARCSEK, Z., MINAROVITS-KORMUTA, S., KLEIN, G. & ERNBERG, I. 1992. RNA polymerase III-transcribed EBER 1 and 2 transcription units are expressed and hypomethylated in the major Epstein-Barr virus-carrying cell types. *J Gen Virol,* 73 (Pt 7), 1687-92.

MOGHADDAM, A., ROSENZWEIG, M., LEE-PARRITZ, D., ANNIS, B., JOHNSON, R. P. & WANG, F. 1997. An animal model for acute and persistent Epstein-Barr virus infection. *Science,* 276, 2030-3.

MORGAN, A. J., ALLISON, A. C., FINERTY, S., SCULLION, F. T., BYARS, N. E. & EPSTEIN, M. A. 1989. Validation of a first-generation Epstein-Barr virus vaccine preparation suitable for human use. *J Med Virol,* 29, 74-8.

MORGAN, A. J., FINERTY, S., LOVGREN, K., SCULLION, F. T. & MOREIN, B. 1988a. Prevention of Epstein-Barr (EB) virus-induced lymphoma in cottontop tamarins by vaccination with the EB virus envelope glycoprotein gp340 incorporated into immune-stimulating complexes. *J Gen Virol,* 69 (Pt 8), 2093-6.

MORGAN, A. J., MACKETT, M., FINERTY, S., ARRAND, J. R., SCULLION, F. T. & EPSTEIN, M. A. 1988b. Recombinant vaccinia virus expressing Epstein-Barr virus glycoprotein gp340 protects cottontop tamarins against EB virus-induced malignant lymphomas. *J Med Virol,* 25, 189-95.

MOTZ, M., DEBY, G. & WOLF, H. 1987. Truncated versions of the two major Epstein-Barr viral glycoproteins (gp250/350) are secreted by recombinant Chinese hamster ovary cells. *Gene,* 58, 149-54.

MOUTSCHEN, M., LEONARD, P., SOKAL, E. M., SMETS, F., HAUMONT, M., MAZZU, P., BOLLEN, A., DENAMUR, F., PEETERS, P., DUBIN, G. & DENIS, M. 2007. Phase I/II studies to evaluate safety and immunogenicity of a recombinant gp350 Epstein-Barr virus vaccine in healthy adults. *Vaccine,* 25, 4697-705.

NAKAGOMI, H., DOLCETTI, R., BEJARANO, M. T., PISA, P., KIESSLING, R. & MASUCCI, M. G. 1994. The Epstein-Barr virus latent membrane protein-1 (LMP1) induces interleukin-10 production in Burkitt lymphoma lines. *Int J Cancer,* 57, 240-4.

NANBO, A., INOUE, K., ADACHI-TAKASAWA, K. & TAKADA, K. 2002. Epstein-Barr virus RNA confers resistance to interferon-alpha-induced apoptosis in Burkitt's lymphoma. *EMBO J,* 21, 954-65.

NEMEROW, G. R., MOLD, C., SCHWEND, V. K., TOLLEFSON, V. & COOPER, N. R. 1987. Identification of gp350 as the viral glycoprotein mediating attachment of Epstein-Barr virus (EBV) to the EBV/C3d receptor of B cells: sequence homology of gp350 and C3 complement fragment C3d. *J Virol,* 61, 1416-20.

NONKWELO, C., SKINNER, J., BELL, A., RICKINSON, A. & SAMPLE, J. 1996. Transcription start sites downstream of the Epstein-Barr virus (EBV) Fp promoter in early-passage Burkitt lymphoma cells define a fourth promoter for expression of the EBV EBNA-1 protein. *J Virol,* 70, 623-7.

NORTH, J. R., MORGAN, A. J., THOMPSON, J. L. & EPSTEIN, M. A. 1982. Purified Epstein-Barr virus Mr 340,000 glycoprotein induces potent virus-neutralizing antibodies when incorporated in liposomes. *Proc Natl Acad Sci U S A,* 79, 7504-8.

OWEN, T. J., O'NEIL, J. D., DAWSON, C. W., HU, C., CHEN, X., YAO, Y., WOOD, V. H., MITCHELL, L. E., WHITE, R. J., YOUNG, L. S. & ARRAND, J. R. 2010. Epstein-Barr virus-encoded EBNA1 enhances RNA polymerase III-dependent EBER expression through induction of EBER-associated cellular transcription factors. *Mol Cancer,* 9, 241.

PIPERI, E., OMLIE, J., KOUTLAS, I. G. & PAMBUCCIAN, S. 2010. Oral hairy leukoplakia in HIV-negative patients: report of 10 cases. *Int J Surg Pathol,* 18, 177-83.

POLACK, A., HORTNAGEL, K., PAJIC, A., CHRISTOPH, B., BAIER, B., FALK, M., MAUTNER, J., GELTINGER, C., BORNKAMM, G. W. & KEMPKES, B. 1996. c-myc activation renders proliferation of Epstein-Barr virus (EBV)-transformed cells independent of EBV nuclear antigen 2 and latent membrane protein 1. *Proc Natl Acad Sci U S A,* 93, 10411-6.

RABBITTS, T. H., BAER, R., DAVIS, M., FORSTER, A., HAMLYN, P. H. & MALCOLM, S. 1984a. The c-myc gene paradox in Burkitt's lymphoma chromosomal translocation. *Curr Top Microbiol Immunol,* 113, 166-71.

RABBITTS, T. H., FORSTER, A., HAMLYN, P. & BAER, R. 1984b. Effect of somatic mutation within translocated c-myc genes in Burkitt's lymphoma. *Nature,* 309, 592-7.

RAGOT, T., FINERTY, S., WATKINS, P. E., PERRICAUDET, M. & MORGAN, A. J. 1993. Replication-defective recombinant adenovirus expressing the Epstein-Barr virus (EBV) envelope glycoprotein gp340/220 induces protective immunity against EBV-induced lymphomas in the cottontop tamarin. *J Gen Virol,* 74 (Pt 3), 501-7.

RAWLINS, D. R., MILMAN, G., HAYWARD, S. D. & HAYWARD, G. S. 1985. Sequence-specific DNA binding of the Epstein-Barr virus nuclear antigen (EBNA-1) to clustered sites in the plasmid maintenance region. *Cell,* 42, 859-68.

RICKINSON, A. B., YOUNG, L. S. & ROWE, M. 1987. Influence of the Epstein-Barr virus nuclear antigen EBNA 2 on the growth phenotype of virus-transformed B cells. *J Virol,* 61, 1310-7.

ROBERTSON, E. S., GROSSMAN, S., JOHANNSEN, E., MILLER, C., LIN, J., TOMKINSON, B. & KIEFF, E. 1995. Epstein-Barr virus nuclear protein 3C modulates transcription through interaction with the sequence-specific DNA-binding protein J kappa. *J Virol,* 69, 3108-16.

ROSA, M. D., GOTTLIEB, E., LERNER, M. R. & STEITZ, J. A. 1981. Striking similarities are exhibited by two small Epstein-Barr virus-encoded ribonucleic acids and the adenovirus-associated ribonucleic acids VAI and VAII. *Mol Cell Biol,* 1, 785-96.

ROWE, M., LEAR, A. L., CROOM-CARTER, D., DAVIES, A. H. & RICKINSON, A. B. 1992. Three pathways of Epstein-Barr virus gene activation from EBNA1-positive latency in B lymphocytes. *J Virol,* 66, 122-31.

ROWE, M., YOUNG, L. S., CADWALLADER, K., PETTI, L., KIEFF, E. & RICKINSON, A. B. 1989. Distinction between Epstein-Barr virus type A (EBNA 2A) and type B (EBNA 2B) isolates extends to the EBNA 3 family of nuclear proteins. *J Virol,* 63, 1031-9.

SAMANTA, M., IWAKIRI, D. & TAKADA, K. 2008. Epstein-Barr virus-encoded small RNA induces IL-10 through RIG-I-mediated IRF-3 signaling. *Oncogene,* 27, 4150-60.

SAMPLE, J., YOUNG, L., MARTIN, B., CHATMAN, T., KIEFF, E. & RICKINSON, A. 1990. Epstein-Barr virus types 1 and 2 differ in their EBNA-3A, EBNA-3B, and EBNA-3C genes. *J Virol,* 64, 4084-92.

SCHOLLE, F., BENDT, K. M. & RAAB-TRAUB, N. 2000. Epstein-Barr virus LMP2A transforms epithelial cells, inhibits cell differentiation, and activates Akt. *J Virol,* 74, 10681-9.

SCHULTZ, L. D., TANNER, J., HOFMANN, K. J., EMINI, E. A., CONDRA, J. H., JONES, R. E., KIEFF, E. & ELLIS, R. W. 1987. Expression and secretion in yeast of a 400-kDa envelope glycoprotein derived from Epstein-Barr virus. *Gene,* 54, 113-23.

SHAKNOVICH, R., BASSO, K., BHAGAT, G., MANSUKHANI, M., HATZIVASSILIOU, G., MURTY, V. V., BUETTNER, M., NIEDOBITEK, G., ALOBEID, B. & CATTORETTI, G. 2006. Identification of rare Epstein-Barr virus infected memory B cells and plasma cells in non-monomorphic post-transplant lymphoproliferative disorders and the signature of viral signaling. *Haematologica,* 91, 1313-20.

SHELDON, P. J., HEMSTED, E. H., PAPAMICHAIL, M. & HOLBOROW, E. J. 1973. Thymic origin of atypical lymphoid cells in infectious mononucleosis. *Lancet,* 1, 1153-5.

SHIBATA, D. & WEISS, L. M. 1992. Epstein-Barr virus-associated gastric adenocarcinoma. *Am J Pathol,* 140, 769-74.

SHIBATA, D., WEISS, L. M., NATHWANI, B. N., BRYNES, R. K. & LEVINE, A. M. 1991. Epstein-Barr virus in benign lymph node biopsies from individuals infected with the human immunodeficiency virus is associated with concurrent or subsequent development of non-Hodgkin's lymphoma. *Blood,* 77, 1527-33.

SINCLAIR, A. J. & FARRELL, P. J. 1992. Epstein-Barr virus transcription factors. *Cell Growth Differ,* 3, 557-63.

SINCLAIR, A. J., PALMERO, I., PETERS, G. & FARRELL, P. J. 1994. EBNA-2 and EBNA-LP cooperate to cause G0 to G1 transition during immortalization of resting human B lymphocytes by Epstein-Barr virus. *EMBO J,* 13, 3321-8.

SISTA, N. D., BARRY, C., SAMPSON, K. & PAGANO, J. 1995. Physical and functional interaction of the Epstein-Barr virus BZLF1 transactivator with the retinoic acid receptors RAR alpha and RXR alpha. *Nucleic Acids Res,* 23, 1729-36.

SIXBEY, J. W., LEMON, S. M. & PAGANO, J. S. 1986. A second site for Epstein-Barr virus shedding: the uterine cervix. *Lancet,* 2, 1122-4.

SIXBEY, J. W., NEDRUD, J. G., RAAB-TRAUB, N., HANES, R. A. & PAGANO, J. S. 1984. Epstein-Barr virus replication in oropharyngeal epithelial cells. *N Engl J Med,* 310, 1225-30.

SLOBEDMAN, B., BARRY, P. A., SPENCER, J. V., AVDIC, S. & ABENDROTH, A. 2009. Virus-encoded homologs of cellular interleukin-10 and their control of host immune function. *J Virol,* 83, 9618-29.

SNUDDEN, D. K., HEARING, J., SMITH, P. R., GRASSER, F. A. & GRIFFIN, B. E. 1994. EBNA-1, the major nuclear antigen of Epstein-Barr virus, resembles 'RGG' RNA binding proteins. *EMBO J,* 13, 4840-7.

SOKAL, E. M., HOPPENBROUWERS, K., VANDERMEULEN, C., MOUTSCHEN, M., LEONARD, P., MOREELS, A., HAUMONT, M., BOLLEN, A., SMETS, F. & DENIS, M. 2007. Recombinant gp350 vaccine for infectious mononucleosis: a phase 2, randomized, double-blind, placebo-controlled trial to evaluate the safety, immunogenicity, and efficacy of an Epstein-Barr virus vaccine in healthy young adults. *J Infect Dis,* 196, 1749-53.

SPENCER, J. V., LOCKRIDGE, K. M., BARRY, P. A., LIN, G., TSANG, M., PENFOLD, M. E. & SCHALL, T. J. 2002. Potent immunosuppressive activities of cytomegalovirus-encoded interleukin-10. *J Virol,* 76, 1285-92.

SWAMINATHAN, S., TOMKINSON, B. & KIEFF, E. 1991. Recombinant Epstein-Barr virus with small RNA (EBER) genes deleted transforms lymphocytes and replicates *in vitro*. *Proc Natl Acad Sci U S A,* 88, 1546-50.

TAKADA, K. 1984. Cross-linking of cell surface immunoglobulins induces Epstein-Barr virus in Burkitt lymphoma lines. *Int J Cancer,* 33, 27-32.

TANNER, J., WEIS, J., FEARON, D., WHANG, Y. & KIEFF, E. 1987. Epstein-Barr virus gp350/220 binding to the B lymphocyte C3d receptor mediates adsorption, capping, and endocytosis. *Cell,* 50, 203-13.

THORLEY-LAWSON, D. A. & POODRY, C. A. 1982. Identification and isolation of the main component (gp350-gp220) of Epstein-Barr virus responsible for generating neutralizing antibodies *in vivo*. *J Virol,* 43, 730-6.

TOCZYSKI, D. P., MATERA, A. G., WARD, D. C. & STEITZ, J. A. 1994. The Epstein-Barr virus (EBV) small RNA EBER1 binds and relocalizes ribosomal protein L22 in EBV-infected human B lymphocytes. *Proc Natl Acad Sci U S A,* 91, 3463-7.

TOCZYSKI, D. P. & STEITZ, J. A. 1991. EAP, a highly conserved cellular protein associated with Epstein-Barr virus small RNAs (EBERs). *EMBO J,* 10, 459-66.

TOKUNAGA, M., LAND, C. E., UEMURA, Y., TOKUDOME, T., TANAKA, S. & SATO, E. 1993. Epstein-Barr virus in gastric carcinoma. *Am J Pathol,* 143, 1250-4.

TOMKINSON, B. & KIEFF, E. 1992. Use of second-site homologous recombination to demonstrate that Epstein-Barr virus nuclear protein 3B is not important for lymphocyte infection or growth transformation *in vitro*. *J Virol,* 66, 2893-903.

TOMKINSON, B., ROBERTSON, E. & KIEFF, E. 1993. Epstein-Barr virus nuclear proteins EBNA-3A and EBNA-3C are essential for B-lymphocyte growth transformation. *J Virol,* 67, 2014-25.

TOVEY, M. G., LENOIR, G. & BEGON-LOURS, J. 1978. Activation of latent Epstein-Barr virus by antibody to human IgM. *Nature,* 276, 270-2.

TSURUMI, T., FUJITA, M. & KUDOH, A. 2005. Latent and lytic Epstein-Barr virus replication strategies. *Rev Med Virol,* 15, 3-15.

VERNINO, S., SALOMAO, D. R., HABERMANN, T. M. & O'NEILL, B. P. 2005. Primary CNS lymphoma complicating treatment of myasthenia gravis with mycophenolate mofetil. *Neurology,* 65, 639-41.

WAGNER, H. J., KLUTER, H., KRUSE, A. & KIRCHNER, H. 1994. [Relevance of transmission of Epstein-Barr virus through blood transfusion]. *Beitr Infusionsther Transfusionsmed,* 32, 138-41.

WANG, D., LIEBOWITZ, D. & KIEFF, E. 1985. An EBV membrane protein expressed in immortalized lymphocytes transforms established rodent cells. *Cell,* 43, 831-40.

WANG, F., GREGORY, C., SAMPLE, C., ROWE, M., LIEBOWITZ, D., MURRAY, R., RICKINSON, A. & KIEFF, E. 1990a. Epstein-Barr virus latent membrane protein (LMP1) and nuclear proteins 2 and 3C are effectors of phenotypic changes in B lymphocytes: EBNA-2 and LMP1 cooperatively induce CD23. *J Virol,* 64, 2309-18.

WANG, F., GREGORY, C. D., ROWE, M., RICKINSON, A. B., WANG, D., BIRKENBACH, M., KIKUTANI, H., KISHIMOTO, T. & KIEFF, E. 1987. Epstein-Barr virus nuclear antigen 2 specifically induces expression of the B-cell activation antigen CD23. *Proc Natl Acad Sci U S A,* 84, 3452-6.

WANG, F., KIKUTANI, H., TSANG, S. F., KISHIMOTO, T. & KIEFF, E. 1991. Epstein-Barr virus nuclear protein 2 transactivates a cis-acting CD23 DNA element. *J Virol,* 65, 4101-6.

WANG, F., TSANG, S. F., KURILLA, M. G., COHEN, J. I. & KIEFF, E. 1990b. Epstein-Barr virus nuclear antigen 2 transactivates latent membrane protein LMP1. *J Virol,* 64, 3407-16.

WEISS, L. M. 2000. Epstein-Barr virus and Hodgkin's disease. *Curr Oncol Rep,* 2, 199-204.

WEISS, L. M., MOVAHED, L. A., WARNKE, R. A. & SKLAR, J. 1989. Detection of Epstein-Barr viral genomes in Reed-Sternberg cells of Hodgkin's disease. *N Engl J Med,* 320, 502-6.

WHANG, Y., SILBERKLANG, M., MORGAN, A., MUNSHI, S., LENNY, A. B., ELLIS, R. W. & KIEFF, E. 1987. Expression of the Epstein-Barr virus gp350/220 gene in rodent and primate cells. *J Virol,* 61, 1796-807.

WILSON, J. B., BELL, J. L. & LEVINE, A. J. 1996. Expression of Epstein-Barr virus nuclear antigen-1 induces B cell neoplasia in transgenic mice. *EMBO J,* 15, 3117-26.

WOISETSCHLAEGER, M., JIN, X. W., YANDAVA, C. N., FURMANSKI, L. A., STROMINGER, J. L. & SPECK, S. H. 1991. Role for the Epstein-Barr virus nuclear antigen 2 in viral promoter switching during initial stages of infection. *Proc Natl Acad Sci U S A,* 88, 3942-6.

YATES, J. L., WARREN, N. & SUGDEN, B. 1985. Stable replication of plasmids derived from Epstein-Barr virus in various mammalian cells. *Nature,* 313, 812-5.

YOSHIYAMA, H., SHIMIZU, N. & TAKADA, K. 1995. Persistent Epstein-Barr virus infection in a human T-cell line: unique program of latent virus expression. *EMBO J,* 14, 3706-11.

YOUNG, L., ALFIERI, C., HENNESSY, K., EVANS, H., O'HARA, C., ANDERSON, K. C., RITZ, J., SHAPIRO, R. S., RICKINSON, A., KIEFF, E. & ET AL. 1989. Expression of Epstein-Barr virus transformation-associated genes in tissues of patients with EBV lymphoproliferative disease. *N Engl J Med,* 321, 1080-5.

YOUNG, L. S. & RICKINSON, A. B. 2004. Epstein-Barr virus: 40 years on. *Nat Rev Cancer,* 4, 757-68.

YOUNG, L. S., YAO, Q. Y., ROONEY, C. M., SCULLEY, T. B., MOSS, D. J., RUPANI, H., LAUX, G., BORNKAMM, G. W. & RICKINSON, A. B. 1987. New type B isolates of Epstein-Barr virus from Burkitt's lymphoma and from normal individuals in endemic areas. *J Gen Virol,* 68 (Pt 11), 2853-62.

YU, M. C., HO, J. H., LAI, S. H. & HENDERSON, B. E. 1986. Cantonese-style salted fish as a cause of nasopharyngeal carcinoma: report of a case-control study in Hong Kong. *Cancer Res,* 46, 956-61.

YU, M. C. & YUAN, J. M. 2002. Epidemiology of nasopharyngeal carcinoma. *Semin Cancer Biol,* 12, 421-9.

ZEIDLER, R., EISSNER, G., MEISSNER, P., UEBEL, S., TAMPE, R., LAZIS, S. & HAMMERSCHMIDT, W. 1997. Downregulation of TAP1 in B lymphocytes by cellular and Epstein-Barr virus-encoded interleukin-10. *Blood,* 90, 2390-7.

ZENG, Y. 1985. Seroepidemiological studies on nasopharyngeal carcinoma in China. *Adv Cancer Res,* 44, 121-38.

ZIMBER-STROBL, U., SUENTZENICH, K. O., LAUX, G., EICK, D., CORDIER, M., CALENDER, A., BILLAUD, M., LENOIR, G. M. & BORNKAMM, G. W. 1991. Epstein-Barr virus nuclear antigen 2 activates transcription of the terminal protein gene. *J Virol,* 65, 415-23.

ZIMMERMANN, H. & TRAPPE, R. U. 2013. EBV and posttransplantation lymphoproliferative disease: what to do? *Hematology Am Soc Hematol Educ Program,* 2013**,** 95-102.

ZUR HAUSEN, H., O'NEILL, F. J., FREESE, U. K. & HECKER, E. 1978. Persisting oncogenic herpesvirus induced by the tumour promotor TPA. *Nature,* 272**,** 373-5.

CHAPTER 7

Human Immunodeficiency Virus

Liljana Stevceva[*]

California Northstate University College of Medicine, 9700 West Taron Drive, Room 138, Elk Grove, CA 95757, USA

Abstract: Human Immunodeficiency Virus (HIV) is a retrovirus that establishes latent infection in humans. The latency period varies in duration but eventually ends in depletion of CD4+ T lymphocytes, secondary immunodeficiency and death from opportunistic infections. The virus is a RNA virus with envelope and glycoprotein spikes protruding from it. The envelope glycoprotein plays an essential role in viral entry into the host cell. In addition, it is an important factor in the capacity of the virus to escape the immune system forming the so called 'glucan shield' that protects the virus from destruction. Both dendritic cells and B cells can carry the infectious virus on their surface as they travel to T lymphocyte rich lymphoid organs and spread the infection through their interaction with T cells. HIV has caused a worldwide epidemic that is currently kept under control with retroviral therapy in the developed world but has devastated large portion of the African continent. Tremendous efforts were put forward by the world research community to develop vaccine against HIV that resulted in moderate to poor protective efficacy of vaccine candidates.

Keywords: HIV, AIDS, Gag, Pol, Env, gp120, CD4+, retrovirus, Vpu, Vif, Vpr, Nef, Tat, Rev, epidemic, latent, virion, budding, escape, immunosupression, glycan shield, HERVs, V1/V2 loop, V3 loop, V4 loop, V5 loop, bridging sheet, gp41, viremia, viral load, depletion of CD4, DC-SIGN, glycosylation, polyclonal activation, complement, seroconversion, CCR5, CXCR4, Tregs

INTRODUCTION

Acquired Immunodeficiency Syndrome (AIDS) was firstly described among homosexual men in New York City and San Francisco in 1981 with reports of unexplained opportunistic infections such as *Pneumocystis jirovecii* pneumonia and Kaposi's sarcoma (Masur *et al.*, 1981), (Durack, 1981), (Gottlieb *et al.*, 1981). It was then discovered that infection with the Human Immunodeficiency virus (HIV) causes AIDS (Barre-Sinoussi *et al.*, 1983).

*Corresponding author Liljana Stevceva: University of Texas Rio Grande Valley School of Medicine, 2102 Treasure Hills Blvd., Harlingen TX, USA; E-mail: liljana@hotmail.com

HIV and AIDS cause a major modern age epidemic with 36.9 million people living with HIV infection worldwide at the end of 2014 (WHO statistics from July, 2015). Of those, 3.2 million are children and 16 million are women. A total of 1.2 million people died of HIV – related causes globally. Currently, this epidemic is especially affecting Sub-Saharan Africa with 25.8 million people living with HIV infection there in 2014 (WHO, 2015).

STRUCTURE OF HIV

The HIV is a retrovirus consisting of an inner core that contains the ribonucleic acid (RNA) genome and the enzymes that are essential for the survival of the virus: the protease that cleaves the viral proteins, the reverse transcriptase that transcribes the RNA back to deoxyribonucleic acid (DNA) once the virus infects a host cell and the integrase that integrates the viral (DNA) into the host cell DNA. The genome size of the HIV is 7-13 kb and contains identical single stranded RNA (+). A capsid protein, p24 and an outer envelope protect the RNA genome.

Envelope

The outer envelope consists of a bi-lipid layer with about 14 glycoprotein spikes. The glycoprotein spikes (gp120) are anchored to the viral membrane as trimers *via* the gp41 protein forming the gp160 complex (Zhu *et al.*, 2006); (Allan *et al.*, 1985); (Robey *et al.*, 1985); (Dowbenko *et al.*, 1988), (Leonard *et al.*, 1990) (see Fig. **1**).

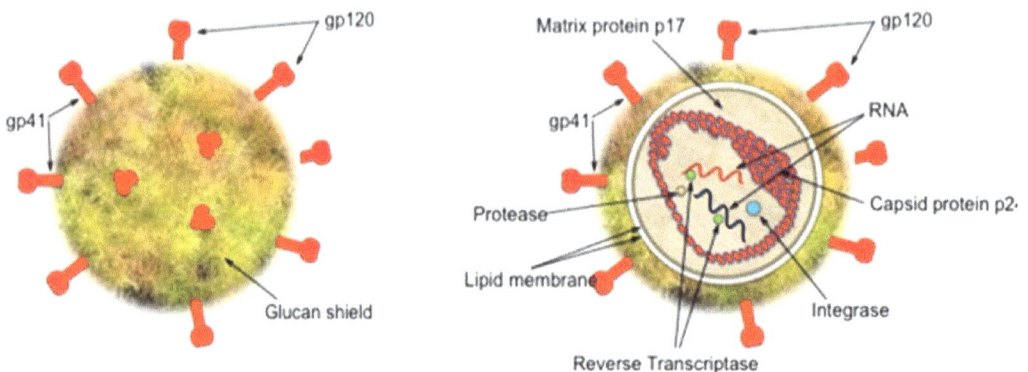

Fig. (1). Structure of the HIV virus. The glycoproteins are oligomannose glycans forming the 'glycan shield' (yellow). The protruding glycoprotein spikes (gp120) are anchored to the envelope with the gp41 protein forming the gp160 complex.

During the budding of the virus from virus-producing cells envelope glycoproteins trimerize and are modified by N-linked glycosylation (Wyatt *et al.*, 1998), (Mao *et al.*, 2012). The carbohydrates were shown to consist predominantly of oligomannose glycans that are attached to envelope proteins (Bonomelli *et al.*, 2011) forming a so called 'glycan shield'.

The glycoprotein spikes are essential for the infectivity of the virus as the virus uses them to attach to the host cell membrane. This predominantly occurs by binding of gp120 to the CD4 receptor on host cells (Dalgleish *et al.*, 1984), (Klatzmann *et al.*, 1984), (Sattentau *et al.*, 1986). The gp120 binding to the CD4 receptor and a co-receptor (CCR5 or CXCR4) is required for fusion of viral membrane with the membrane of the host cell (Moser, 1998), (Wu *et al.*, 1996) (see Fig. **2**).

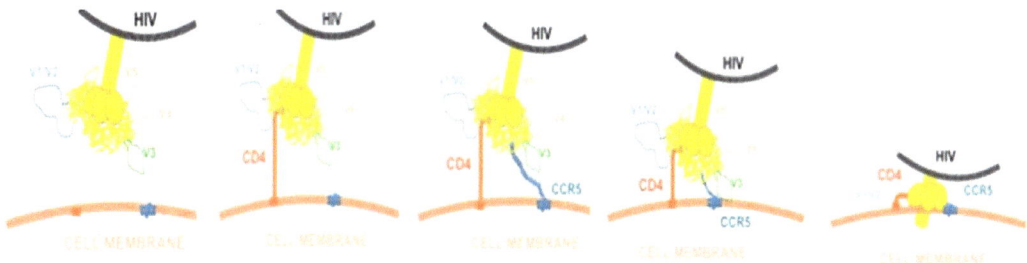

Fig. (2). Attachment of HIV to the host cell. The virus attaches to the CD4 host cell receptor and to a co-receptor (mostly CCR5 but also CXCR4) that pull it through the cell membrane into the cytoplasm.

Because of its importance, the structure of the gp120 glycoprotein has been investigated extensively. While the gp41 portion does not contain any highly variable regions (about 80% of amino acids are highly conserved) the gp120 portion consists of five conserved regions (C1-C5) and five variable loops (V1-V5) (Wyatt and Sodroski, 1998); (Modrow *et al.*, 1987). The variable loops are anchored at their bases with disulfide bonds (Leonard *et al.*, 1990). There are two main domains of gp120: the inner, more conserved domain consisting of V1 and V2 loops and the outer domain consisting of V3, V4 and V5. The inner and the outer domain interface with the bridging sheet. The bridging sheet is an area of the gp120 molecule that is involved in the binding of a co-receptor of CD4, the receptor for the chemokine CCR5 (Liu *et al.*, 2003). This particular region is highly conserved in all of HIV viruses and it is thought that the co-receptor affinity may be linked to viral tropism and subsequently, to pathogenesis. On the native gp120 most of the conserved regions are not accessible to neutralizing antibodies (Moore *et al.*, 1994). Only parts of V2 and V3 are exposed on the oligomeric gp120 but the base of V3 is not exposed. In addition, most of V5 and a portion of V4 are very poorly accessible to monoclonal antibodies. Therefore, there are very few well conserved epitopes on the gp120 surface that are exposed and vulnerable to binding and destruction by antibodies. In addition, when under immunological pressure, the variable loops can mutate and destroy the exposed epitopes (Moore *et al*., 1994).

The elegant studies by Leaonard *et al.* have shown that the Gp120 has 24 potential N-glycosylation sites, and that it contains 18 cysteine residues and 9 disulfide bonds (Leonard *et al*., 1990). Of those, three are located on the V1/V2 loops connecting them to each other. The gp120's polypeptide core has 60 000 daltons. The carbohydrate portion adds as much to the total molecular weight bringing it to 120 000 daltons (Lasky *et al*., 1986). Most of the carbohydrate content of the molecule seems to belong to the V1/V2 loop as deletion of the V1/V2 loop leads to significant reduction of the carbohydrate content of the gp120 molecule (Reitter *et al*., 1998). An increase of the positive charge in V1/V2 loop in late infection is accompanied with a switch in co-receptor use for viral entry from CCR5 to CxCR4 (Wang *et al*, 1995; Cornelissen *et al*., 1995; Groenink *et al*., 1993). It is believed that the carbohydrate component of gp120 has a protective function and that it shields the virus from recognition by anti-viral antibodies. A recent study by Reitter *et al.* (Reitter *et al*., 1998) has provided some evidence that this might be true at least for the V1 region. In a series of experiments using a derivative of SIVmac239 lacking N-linked sites 5, 6, 8, 12 and 13, the absence of two N-linked glycosylation sites around/in V1 loop dramatically increased neutralizing antibody responses. Above all, the conserved regions of gp120 are extensively shielded by the carbohydrates (Reitter *et al*., 1998) and the co-receptor binding site is hidden by the variable loops until binding to CD4 receptor occurs (Kwong *et al*., 1998, Wyatt *et al*., 1993).

The five variable loops of gp120 are essential in the process of viral entry and assembly. The V1/V2 loop consists of one loop of about 25 amino acids for V1 and about 45 amino acids for V2 (Wyatt *et al*., 1998). It has been shown that deletion of V1 and V2 abolishes virus entry. The V3 loop consists of about 30 amino acids and it is considered to be determining the tropism of the virus as well as the co-receptor use (Yang *et al*., 2004). Deletion of V3 also abolished viral entry while deletion of V3 crown significantly enhanced virus assembly and entry. Deletion of V4 or V5 knocked out the receptor and co-receptor binding sites in gp120, and negatively affected envelope (env) cell surface display, leading to the failure in virus assembly and subsequent entry (Yuan *et al*., 2013).

It has been postulated that the variable loops change their structure and conformation and this is an important factor in the virus ability to elude the immune response. This was shown to be especially true for the V3 loop (Schreiber *et al.*, 1997, Yuan *et al.*, 2013, Kliks *et al.*, 1993). Similarly, a study done by Schreiber *et al.* had shown lack of V3 specific neutralizing antibodies in patients that develop AIDS (Schreiber *et al.*, 1997). In addition, neutralizing antibodies for one HIV strain to the V3 loop fail to neutralize another strain of the same virus and can enhance infection in the third strain (Kliks *et al.*, 1993).

The V3 loop of the gp120 molecule also determines the chemokine co-receptor usage (Cocchi *et al.*, 1996), (Speck *et al.*, 1997). Early infection utilizes CCR5 as a co receptor and later infection CXCR4 (Choe *et al.*, 1996), (Connor *et al.*, 1997), (Zhang *et al.*, 1996). When CD4 binds to V1/V2 region, V1 and V2 shift to expose residues that are otherwise protected (Morikita *et al.*, 1997) presenting the necessary region for co-receptor binding with CXCR4 (Kwong *et al.*, 1998) or CCR5 (Alkhatib *et al.*, 1996).

Structural Proteins

The Gag protein is one of the main components of the virus making about 50% of the virus, lipid membrane is about 30% of the virion, other viral proteins are about 20% and RNA is about 2.5% (reviewed in Carlson *et al.* (Wright *et al.*, 2007, Carlson *et al.*, 2008). The *Gag* protein is organized in hexamers with 8nm spacing creating a lattice (Wright *et al.*, 2007). Assembly of the HIV virus is guided and coordinated by the immature *Gag* polyprotein. Following budding, proteolytic degradation of *Gag* results in a formation of the characteristic conical capsid. As a matter of fact, the viral protease cleaves the immature *Gag* into six proteins: matrix protein (p17 or MA), capsid protein (p24 or CA), nucleocapsid (NC), p6, SP2 (p2) and SP1 (p1) (Ganser-Pornillos *et al.*, 2008), (Takemura and Murakami, 2012), (Sundquist and Krausslich, 2012). This proteolysis is required for the virus to become infectious. The resulting capsid protein forms a conical structure encapsulating two RNA molecules connected by nucleocapsid proteins (Takemura and Murakami, 2012).

The Pol polyprotein consists of the viral enzymes **reverse transcriptase** that transcribes DNA from RNA, **integrase** that integrates viral DNA into the host genome and **protease** that cleaves the *Gag* polyprotein (Haseltine, 1991).

Regulatory Proteins

Important regulatory proteins of HIV are ***Tat protein*** that regulates the reverse transcription and ***Rev*** that regulates viral protein expression. Accessory regulatory proteins are *Vif* (important for infectivity), ***Vpu*** (involved in CD4 degradation and release of virions from infected cells), ***Vpr*** (important for viral replication) and *Nef* (important in viral infectivity and cell apoptosis).

After the budding, the HIV virus acquires its lipid membrane from the infected cell and releases itself. This process is essential for the spreading of the viral infection. The antiviral protein tetherin blocks the release of the newly formed HIV virion and its spread by tethering it to the cell surface. This molecule is antagonized by HIV Vpu protein (Malim and Bieniasz, 2012). Absence of Vpu traps the nascent new virions to the cell surface and inhibits release (Neil *et al.*, 2007).

THE DYNAMIC OF THE HIV INFECTION

In most of cases HIV infection occurs by sexual transmission (70-80%) of which 60-70% occur *via* vaginal intercourse and 5-10% *via* anal intercourse. The second most common route is the parenteral infection (8-15%) that occurs in intravenous drug users by contaminated needle sharing, accidental needle stick in health care workers and contaminated blood products. The rest of the HIV infections are attributed to perinatal infections (5-10%) infecting the unborn child *in utero*, during labor or even after delivery through breastfeeding.

Immediately after exposure, viral replication occurs in the lymphoid tissue draining the inoculation sites. Viremia occurs 1-3 weeks after infection and can be detected by p24 antigen or RNA assays. The majority of patients seroconvert within 8 weeks after the primary infection but about 5% have delayed seroconversion window of more than 6 months after exposure. The neutralizing antibodies appear in the peripheral blood and the HIV infection can be diagnosed by using Enzyme Linked Immunosorbent Assay (ELISA) or Western blot for HIV-specific antibodies (Busch and Satten, 1997). The third generation of ELISA tests can detect HIV antibodies as early as week 3 post-infection (Ly *et al.*, 2001). The new fourth generation tests that are a combination of (ELISA) with p24 antigen shorten the time for detecting HIV infection to 15 days and improve sensitivity and specificity (Ly *et al.*, 2001), (Sickinger *et al.*, 2008). Early detection is likely to decrease the risk of transmission that is much higher in the acute than in established HIV infection (Wawer *et al.*, 2005), (Cohen *et al.*, 2011). The higher risk of transmission is most closely correlated to blood viremia (Cohen *et al.*, 2011).

Cytotoxic Lymphocyte (CTL) responses appear before the peak of viremia and eventually succeed to bring viremia under control but are unable to clear the virus

(Borrow *et al*., 1994), (Koup *et al*., 1994). Because the virus predominantly infects CD4+ cells, this CTL response also leads to direct killing of infected CD4+ cells and a resulting decrease in CD4+ T helper cells numbers. Also, one of the popular hypotheses on the reasons for inefficiency of the immune system to clear HIV is that HIV causes hyperactivation of T cell responses leading to exhaustion and apoptosis (Fauci *et al*., 1996). This hyperactivation was shown to be independent on the viral load (Deeks *et al*., 2004), (Hunt *et al*., 2011) and was attributed to the translocation of microbial products through the disrupted mucosal barrier of the intestine (Douek, 2007) as well as to reactivation of latent infections with cytomegalovirus and Epstein-Barr virus (Doisne *et al*., 2004).

It is now known that massive CD4+ T memory T cell depletion occurs early in the HIV infection but is countered with CD4+ memory T cell regeneration that prevents occurrence of immunodeficiency (Okoye and Picker, 2013). In addition to CTL killing of infected CD4+ T cells, their loss has also been attributed to Fas/Fas ligand-mediated apoptosis (Ameisen and Capron, 1991), (Terai *et al*., 1991). The CD4+ T cells in chronic HIV highly express the programmed death protein (PD-1) (Day *et al*., 2006). Increased numbers of regulatory T cells are also observed. Experiments in SHIV-infected macaques showed that the accumulation of Tregs during acute infection is even greater in lymphoid tissues and that these cells secrete TGFβ upon stimulation with gp120 (Stevceva *et al*., 2008). Because one of the well-known effects of TGFβ is to induce CTL suppression and to promote development of Tregs (Fu *et al*., 2004) this further perpetuates the immunosupression.

Infection with HIV eventually results in depletion of CD4+ T lymphocytes and this is the central event that leads to the immunodeficiency and opportunistic infections that characterize the full-blown acquired immunodeficiency syndrome (AIDS). During the CD4+ T cells destruction predominant target are the memory CD4+ T cells because they are the cells that preferentially express CCR5, the co-receptor used for HIV entry into the cells (Poles *et al*., 2001). This leaves precursor CD4+ T cells intact and capable to generate new memory CD4+ T cells that will allow the virus to continue to multiply (Okoye and Picker, 2013). The constant replenishment of CD4+ T cell population is the reason that acute HIV infection does not immediately results in AIDS. This also suggests that development of AIDS might potentially be avoided if precursor CD4+ T cell pools are somehow kept intact. Eventually, the process of CD4+ T cells renewal fails and this leads to progressive decrease in CD4+ T cell counts in blood. In general, when CD4+ T cell counts decrease to 200/mm3 of blood or less, AIDS-associated opportunistic infections begin to develop.

The antibody response to HIV occurs 2-5 weeks after exposure. This, first antibody response is polyclonal in nature and consist not only of HIV-specific antibodies but also of many other antibodies depending on previous pathogens that the individual was exposed to (Stevceva *et al.*, 2007), (Levesque *et al.*, 2009). This is by in large mediated by the HIV envelope. For example, the carboxy terminus of the gp41 region (amino acids 739-863) induces polyclonal activation of normal B cells, and their differentiation into immunoglobulin-secreting cells (Chirmule *et al.*, 1990). Also, binding of gp120 to the VH3 domain of surface IgM on B cells causes T cell-independent B cell differentiation (Berberian *et al.*, 1993). Although these antibodies are not neutralizing, some of them still bind to the surface of the HIV virus. The bound antibodies amplify the activation of the complement cascade and deposition of complement components to the viral surface (Willey *et al.*, 2011). The deposition of the complement components such as C3a and C5a on the virions facilitates binding of HIV to cells that express complement receptors CR3 and CR4. CR3 (CD11b, CD18) is pattern recognition, cell surface receptor that is mostly found on the surface of macrophages, neutrophils and NK cells and that belongs to the integrin family. CR3 recognizes iC3b component. Binding of an infected cell or HIV to CR3 induces phagocytosis. CR3 is also expressed on the surface of follicular dendritic cells. CR4 (Integrin alphaXbeta2) is also an integrin, composed of CD11c and CD18 that binds to iC3b and is expressed on monocytes, granulocytes, NK cells and some lymphocytes. Binding to CR4 induces phagocytosis and migration of leukocytes (Robert Graham Quinton Leslie, 2001).

Direct activation of the complement cascade by HIV also occurs by binding of the heavily glycosylated gp120 to mannose binding lectin (MBL) and initiating the lectin pathway of complement activation (reviewed in (Ji *et al.*, 2005). The high level of mannose glycosylation on gp120 facilitates this process. Dendritic cells express a C-type lectin present on dendritic cells, called dendritic cell-specific intercellular adhesion molecule 3-grabbing nonintegrin (DC-SIGN). Normally, binding of DC-SIGN results in uptake of pathogens and antigen presentation to T cells. However, binding of HIV to DC-SIGN causes transfer of the HIV infection from DCs to T cells (Geijtenbeek and van Kooyk, 2003). In addition, it has been demonstrated that the heavily glycosylated and partly sialylated envelope proteins gp120 and gp41 bind to cells expressing the complement receptor CR2 (CD21). The binding requires activation of the alternative complement pathway and in the absence of CD4, it does not result in cell infection [68].

Dendritic cells at mucosal entry sites bind HIV *via* their DC-specific C-type lectin, DC-SIGN (Geijtenbeek *et al.*, 2000). This interaction occurs with the gp120 protein on the HIV envelope and it does not result in viral entry into the dendritic cell. Instead, the dendritic cells carry the infectious virus on their surface as they migrate into the mucosal lymphoid tissues. These tissues are very rich in CD4+CCR5+ T cells that are now exposed to the HIV infection *via* the normal Ag presentation process. Gp120 can bind two other receptors in a similar manner, mannose receptor on immature DC and langerin on Langerhans cells (Turville *et al.*, 2002). In addition, follicular dendritic cells located in the germinal centers of all peripheral lymphoid tissues also bind and trap the HIV in immune complexes (Embretson *et al.*, 1993). The immune complexes bound HIV remains infectious for at least 9 months and can infect permissive cells (Smith *et al.*, 2001).

B cells also have the capacity to transfer infective HIV to other, permissive cells such as T cells. The complement opsonized virus binds to the surface of the B cell *via* the CD21 receptor (reviewed in (Stevceva *et al.*, 2007). Therefore, B cells of HIV-1 infected viremic patients carry replication-competent virus on their surface and can infect T cells *via* regular B-T cells interactions (Moir *et al.*, 2000, Jakubik *et al.*, 2000). Such HIV-1 virus that is trapped within the immune complexes and bound to the B cells is more infectious than the free virion (Jakubik *et al.*, 2000).

This initial antibody response to HIV is not only ineffective but it also enhances infection, in some cases up to 350 fold and thus converts a low level infection into a highly destructive one (Willey *et al.*, 2011). By day 47 after transmission, there is damage to the GALT germinal centers and destruction of follicular dendritic cells networks in the ileum Payer's patches (Levesque *et al.*, 2009). Neutralizing HIV antibodies do not appear until, at least, 52 days after appearance of the first HIV-specific antibodies (Wei *et al.*, 2003). Complement-mediated antibody-dependent enhancement of infection declines once the neutralizing antibodies appear but the virus mutates under immunological pressure and later virus isolates that escaped neutralization have greater capacity for enhanced infectivity by autologous antibodies (Willey *et al.*, 2011).

As reviewed above and especially during the acute phase of HIV infection, the gp120 protein elicits many immune responses that contribute to the viral escape. In other words, gp120 has a deleterious effect on the three main players of the immune response: the primary antigen presenting cell, dendritic cell, B cells and T cells. In fact, it would appear that gp120, through its fugitactic activity for T cells, the polyclonal activation of B cells that leads to the formation of a protective coating of complement complexes and through its binding and persistence on FDC processes shields the HIV virus from destruction by the immune mechanisms of the body. The extent and the exact mechanisms by which this occurs are yet to be dissected (Stevceva *et al.*, 2007). Experiments with HIV virus mutated in the gp4l transmembrane envelope glycoprotein markedly reduced the cytopathic effects of HIV on CD4+ lymphocytes pointing out to the envelope as essential component for the cytopathic effect (Kowalski *et al.*, 1991). The processes described above allow spread of the HIV infection to the key players of the immune response and establishment of chronic infection. In other words, the virus 'hijacks' the immune system and utilizes normal immune responses to infect the T cells and to prevent its own lysis.

VACCINE AGAINST HIV INFECTION

In September of 1985 AIDS and WHO Collaborating Centers issued a memorandum from the WHO meeting held in Geneva in which among other things they outlined that the best strategy to fight the spread of AIDS was to prevent the infection by developing a vaccine against HIV. Since then, substantial efforts by the world research community has been dedicated to developing an effective vaccine against HIV. The first HIV vaccine clinical trial opened at the National Institutes of Health (NIH) Clinical Center testing a gp160 subunit vaccine. Now, 30 years later the world is not closer to developing a safe and effective vaccine against HIV.

Several vaccine candidates have made it to Phase III clinical trials. The first large scale HIV vaccine trial began in 1998 when VaxGen initiated Phase III trial of AIDSVAX in North America and the Netherlands. AIDSVAX was a gp120-based vaccine and was designed to induce antibody response against the virus. Because of the refusal of the US Government to support the Phase III trials with AIDVAX, the trials were supported by private funds raised by the developer Genentech. The trials in North America and the Netherlands and a later trial in Thailand showed no difference in infection rates between the immunized group in comparison to the control. Despite this, AIDSVAX was again used as a boost in the RV144 Phase III trial in combination with the ALVAC canarypox-based vaccine. This combination that was tested on 16,402 healthy man and women induced very moderate response with vaccine efficacy of 26.4% in the intention to treat analysis and 26.2% in the per protocol analysis (Rerks-Ngarm *et al.*, 2009).

The efforts of developing efficient protective HIV vaccine are significantly hampered by the fact that really, the immune correlates of protection against the HIV infection are still not known. In a recently published case-control analysis of antibody and cellular immune correlates of infection risk, the binding of IgG antibodies to V1/V2 regions of the HIV-1 envelope protein correlated inversely with the rate of HIV-1 infection and the binding of plasma IgA antibodies correlated directly to the rate of infection. This led to a conclusion that high levels of V1/V2 antibodies may contribute to protection against HIV infection (Haynes *et al.*, 2012).

It has been known for a while that cellular immunity plays an important role in controlling the HIV infection. In a study published in 1999 depletion of CD8+ T cells in HIV infected non-human primates induced substantial increase of viremia and progression of the disease (Schmitz *et al.*, 1999). In addition, studies in elite controllers (HIV infected individuals maintaining HIV RNA to less than 50 copies/ml for at least a year without therapy) have shown that elite control of HIV infection is associated with expression of HLA-B*57, dominant CTL targeting of Gag rather than the Env antigens, lower total magnitude and breadth of HIV-specific CTL response, increased functionality of CD4+ and CD8+ T cells, and weak neutralizing antibodies. Aviremic controllers have statistically significantly higher percentages of CD4+ T helper cells and CTLs that produce both interleukin-2 and interferon gamma (IFN-γ) than do either viremic controllers or progressors (Walker, 2007). These experiments led to the hypothesis that perhaps a better approach is to design a vaccine that will induce virus-specific T cell responses. In trying to induce predominantly T cell responses two major vaccine design approaches have been tried so far. The first one is using viral vectors to deliver the genes of interest. Several viral vectors were used for that purpose including New York Vaccinia virus strain (NYVAC), and Modified Vaccinia Ankara (MVA)] and recombinant adenovirus serotype 5 (rAd5). Of these the rAD5-based vaccine was sufficiently immunogenic to be tested in humans. rAD5 that expressed Gag, Pol and Nef genes was tested in high-risk individuals from North and South America, the Caribbean, and Australia. Unfortunately, vaccine did not show any protective effect and the trial was halted (Buchbinder *et al.*, 2008). Moreover, a higher rate of HIV infections was observed in vaccinees compared with placebo recipients but the difference was not statistically significant. A study that evaluated vaccine-induced immunity demonstrated that the vaccine induced HIV-specific CD4+ T cell responses in only 41% of the vaccines with a median magnitude of 0.2-0.3%. HIV-specific CD8+ T cell responses were elicited at a magnitude of 0.5-1% of circulating CD8+ T cells.

Pre-existing immunity to adenovirus 5 negatively affected the magnitude of the immune response to this vaccine (McElrath *et al.*, 2008).

The second approach to designing T-cell responses inducing vaccines is DNA vaccines. DNA vaccines consist of genetically engineered DNA that is injected directly without using viral vectors. The advantage is that there is no immune response against the vector that could influence the development of the immune response. Unfortunately, DNA vaccines are poorly immunogenic so considerable research effort is now focused on improving the method of delivery and the formulations in order to improve the efficacy of DNA vaccines (Ferraro *et al.*, 2011).

CONFLICT OF INTEREST

The author confirms that this chapter contents have no conflict of interest.

ACKNOWLEDGEMENTS

Declared None.

REFERENCES

ALKHATIB, G., COMBADIERE, C., BRODER, C. C., FENG, Y., KENNEDY, P. E., MURPHY, P. M. & BERGER, E. A. 1996. CC CKR5: a RANTES, MIP-1alpha, MIP-1beta receptor as a fusion cofactor for macrophage-tropic HIV-1. *Science,* 272, 1955-8.

ALLAN, J. S., COLIGAN, J. E., BARIN, F., MCLANE, M. F., SODROSKI, J. G., ROSEN, C. A., HASELTINE, W. A., LEE, T. H. & ESSEX, M. 1985. Major glycoprotein antigens that induce antibodies in AIDS patients are encoded by HTLV-III. *Science,* 228, 1091-4.

AMEISEN, J. C. & CAPRON, A. 1991. Cell dysfunction and depletion in AIDS: the programmed cell death hypothesis. *Immunol Today,* 12, 102-5.

BARRE-SINOUSSI, F., CHERMANN, J. C., REY, F., NUGEYRE, M. T., CHAMARET, S., GRUEST, J., DAUGUET, C., AXLER-BLIN, C., VEZINET-BRUN, F., ROUZIOUX, C., ROZENBAUM, W. & MONTAGNIER, L. 1983. Isolation of a T-lymphotropic retrovirus from a patient at risk for acquired immune deficiency syndrome (AIDS). *Science,* 220, 868-71.

BERBERIAN, L., GOODGLICK, L., KIPPS, T. J. & BRAUN, J. 1993. Immunoglobulin VH3 gene products: natural ligands for HIV gp120. *Science,* 261, 1588-91.

BONOMELLI, C., DOORES, K. J., DUNLOP, D. C., THANEY, V., DWEK, R. A., BURTON, D. R., CRISPIN, M. & SCANLAN, C. N. 2011. The glycan shield of HIV is predominantly oligomannose independently of production system or viral clade. *PLoS One,* 6, e23521.

BORROW, P., LEWICKI, H., HAHN, B. H., SHAW, G. M. & OLDSTONE, M. B. 1994. Virus-specific CD8+ cytotoxic T-lymphocyte activity associated with control of viremia in primary human immunodeficiency virus type 1 infection. *J Virol,* 68, 6103-10.

BUCHBINDER, S. P., MEHROTRA, D. V., DUERR, A., FITZGERALD, D. W., MOGG, R., LI, D., GILBERT, P. B., LAMA, J. R., MARMOR, M., DEL RIO, C., MCELRATH, M. J., CASIMIRO, D. R., GOTTESDIENER, K. M., CHODAKEWITZ, J. A., COREY, L., ROBERTSON, M. N. & STEP STUDY PROTOCOL, T. 2008. Efficacy assessment of a cell-mediated immunity HIV-1 vaccine (the Step Study): a double-blind, randomised, placebo-controlled, test-of-concept trial. *Lancet,* 372, 1881-93.

BUSCH, M. P. & SATTEN, G. A. 1997. Time course of viremia and antibody seroconversion following human immunodeficiency virus exposure. *Am J Med,* 102, 117-24; discussion 125-6.

CARLSON, L. A., BRIGGS, J. A., GLASS, B., RICHES, J. D., SIMON, M. N., JOHNSON, M. C., MULLER, B., GRUNEWALD, K. & KRAUSSLICH, H. G. 2008. Three-dimensional analysis of budding sites and released virus suggests a revised model for HIV-1 morphogenesis. *Cell Host Microbe,* 4, 592-9.

CHIRMULE, N., KALYANARAMAN, V. S., SAXINGER, C., WONG-STAAL, F., GHRAYEB, J. & PAHWA, S. 1990. Localization of B-cell stimulatory activity of HIV-1 to the carboxyl terminus of gp41. *AIDS Res Hum Retroviruses,* 6, 299-305.

CHOE, H., FARZAN, M., SUN, Y., SULLIVAN, N., ROLLINS, B., PONATH, P. D., WU, L., MACKAY, C. R., LAROSA, G., NEWMAN, W., GERARD, N., GERARD, C. & SODROSKI, J. 1996. The beta-chemokine receptors CCR3 and CCR5 facilitate infection by primary HIV-1 isolates. *Cell,* 85, 1135-48.

COCCHI, F., DEVICO, A. L., GARZINO-DEMO, A., CARA, A., GALLO, R. C. & LUSSO, P. 1996. The V3 domain of the HIV-1 gp120 envelope glycoprotein is critical for chemokine-mediated blockade of infection. *Nat Med,* 2, 1244-7.

COHEN, M. S., SHAW, G. M., MCMICHAEL, A. J. & HAYNES, B. F. 2011. Acute HIV-1 Infection. *N Engl J Med,* 364, 1943-54.

CONNOR, R. I., SHERIDAN, K. E., CERADINI, D., CHOE, S. & LANDAU, N. R. 1997. Change in coreceptor use correlates with disease progression in HIV-1--infected individuals. *J Exp Med,* 185, 621-8.

DALGLEISH, A. G., BEVERLEY, P. C., CLAPHAM, P. R., CRAWFORD, D. H., GREAVES, M. F. & WEISS, R. A. 1984. The CD4 (T4) antigen is an essential component of the receptor for the AIDS retrovirus. *Nature,* 312, 763-7.

DAY, C. L., KAUFMANN, D. E., KIEPIELA, P., BROWN, J. A., MOODLEY, E. S., REDDY, S., MACKEY, E. W., MILLER, J. D., LESLIE, A. J., DEPIERRES, C., MNCUBE, Z., DURAISWAMY, J., ZHU, B., EICHBAUM, Q., ALTFELD, M., WHERRY, E. J., COOVADIA, H. M., GOULDER, P. J., KLENERMAN, P., AHMED, R., FREEMAN, G. J. & WALKER, B. D. 2006. PD-1 expression on HIV-specific T cells is associated with T-cell exhaustion and disease progression. *Nature,* 443, 350-4.

DEEKS, S. G., KITCHEN, C. M., LIU, L., GUO, H., GASCON, R., NARVAEZ, A. B., HUNT, P., MARTIN, J. N., KAHN, J. O., LEVY, J., MCGRATH, M. S. & HECHT, F. M. 2004. Immune activation set point during early HIV infection predicts subsequent CD4+ T-cell changes independent of viral load. *Blood,* 104, 942-7.

DOISNE, J. M., URRUTIA, A., LACABARATZ-PORRET, C., GOUJARD, C., MEYER, L., CHAIX, M. L., SINET, M. & VENET, A. 2004. CD8+ T cells specific for EBV, cytomegalovirus, and influenza virus are activated during primary HIV infection. *J Immunol,* 173, 2410-8.

DOUEK, D. 2007. HIV disease progression: immune activation, microbes, and a leaky gut. *Top HIV Med,* 15, 114-7.

DOWBENKO, D., NAKAMURA, G., FENNIE, C., SHIMASAKI, C., RIDDLE, L., HARRIS, R., GREGORY, T. & LASKY, L. 1988. Epitope mapping of the human immunodeficiency virus type 1 gp120 with monoclonal antibodies. *J Virol,* 62, 4703-11.

DURACK, D. T. 1981. Opportunistic infections and Kaposi's sarcoma in homosexual men. *N Engl J Med,* 305, 1465-7.

EMBRETSON, J., ZUPANCIC, M., RIBAS, J. L., BURKE, A., RACZ, P., TENNER-RACZ, K. & HAASE, A. T. 1993. Massive covert infection of helper T lymphocytes and macrophages by HIV during the incubation period of AIDS. *Nature,* 362, 359-62.

FAUCI, A. S., PANTALEO, G., STANLEY, S. & WEISSMAN, D. 1996. Immunopathogenic mechanisms of HIV infection. *Ann Intern Med,* 124, 654-63.

FERRARO, B., MORROW, M. P., HUTNICK, N. A., SHIN, T. H., LUCKE, C. E. & WEINER, D. B. 2011. Clinical applications of DNA vaccines: current progress. *Clin Infect Dis,* 53, 296-302.

FU, S., ZHANG, N., YOPP, A. C., CHEN, D., MAO, M., CHEN, D., ZHANG, H., DING, Y. & BROMBERG, J. S. 2004. TGF-beta induces Foxp3 + T-regulatory cells from CD4 + CD25 - precursors. *Am J Transplant,* 4, 1614-27.

GANSER-PORNILLOS, B. K., YEAGER, M. & SUNDQUIST, W. I. 2008. The structural biology of HIV assembly. *Curr Opin Struct Biol,* 18, 203-17.

GEIJTENBEEK, T. B., KWON, D. S., TORENSMA, R., VAN VLIET, S. J., VAN DUIJNHOVEN, G. C., MIDDEL, J., CORNELISSEN, I. L., NOTTET, H. S., KEWALRAMANI, V. N., LITTMAN, D.

R., FIGDOR, C. G. & VAN KOOYK, Y. 2000. DC-SIGN, a dendritic cell-specific HIV-1-binding protein that enhances trans-infection of T cells. *Cell,* 100, 587-97.

GEIJTENBEEK, T. B. & VAN KOOYK, Y. 2003. DC-SIGN: a novel HIV receptor on DCs that mediates HIV-1 transmission. *Curr Top Microbiol Immunol,* 276, 31-54.

GOTTLIEB, M. S., SCHROFF, R., SCHANKER, H. M., WEISMAN, J. D., FAN, P. T., WOLF, R. A. & SAXON, A. 1981. Pneumocystis carinii pneumonia and mucosal candidiasis in previously healthy homosexual men: evidence of a new acquired cellular immunodeficiency. *N Engl J Med,* 305, 1425-31.

HASELTINE, W. A. 1991. Molecular biology of the human immunodeficiency virus type 1. *FASEB J,* 5, 2349-60.

HAYNES, B. F., GILBERT, P. B., MCELRATH, M. J., ZOLLA-PAZNER, S., TOMARAS, G. D., ALAM, S. M., EVANS, D. T., MONTEFIORI, D. C., KARNASUTA, C., SUTTHENT, R., LIAO, H. X., DEVICO, A. L., LEWIS, G. K., WILLIAMS, C., PINTER, A., FONG, Y., JANES, H., DECAMP, A., HUANG, Y., RAO, M., BILLINGS, E., KARASAVVAS, N., ROBB, M. L., NGAUY, V., DE SOUZA, M. S., PARIS, R., FERRARI, G., BAILER, R. T., SODERBERG, K. A., ANDREWS, C., BERMAN, P. W., FRAHM, N., DE ROSA, S. C., ALPERT, M. D., YATES, N. L., SHEN, X., KOUP, R. A., PITISUTTITHUM, P., KAEWKUNGWAL, J., NITAYAPHAN, S., RERKS-NGARM, S., MICHAEL, J. H. & KIM, J. H. 2012. Immune-correlates analysis of an HIV-1 vaccine efficacy trial. *N Engl J Med,* 366, 1275-86.

HUNT, P. W., CAO, H. L., MUZOORA, C., SSEWANYANA, I., BENNETT, J., EMENYONU, N., KEMBABAZI, A., NEILANDS, T. B., BANGSBERG, D. R., DEEKS, S. G. & MARTIN, J. N. 2011. Impact of CD8+ T-cell activation on CD4+ T-cell recovery and mortality in HIV-infected Ugandans initiating antiretroviral therapy. *AIDS,* 25, 2123-31.

JAKUBIK, J. J., SAIFUDDIN, M., TAKEFMAN, D. M. & SPEAR, G. T. 2000. Immune complexes containing human immunodeficiency virus type 1 primary isolates bind to lymphoid tissue B lymphocytes and are infectious for T lymphocytes. *J Virol,* 74, 552-5.

JI, X., GEWURZ, H. & SPEAR, G. T. 2005. Mannose binding lectin (MBL) and HIV. *Mol Immunol,* 42, 145-52.

KLATZMANN, D., BARRE-SINOUSSI, F., NUGEYRE, M. T., DANQUET, C., VILMER, E., GRISCELLI, C., BRUN-VEZIRET, F., ROUZIOUX, C., GLUCKMAN, J. C., CHERMANN, J. C. & *ET AL.* 1984. Selective tropism of lymphadenopathy associated virus (LAV) for helper-inducer T lymphocytes. *Science,* 225, 59-63.

KLIKS, S. C., SHIODA, T., HAIGWOOD, N. L. & LEVY, J. A. 1993. V3 variability can influence the ability of an antibody to neutralize or enhance infection by diverse strains of human immunodeficiency virus type 1. *Proc Natl Acad Sci U S A,* 90, 11518-22.

KOUP, R. A., SAFRIT, J. T., CAO, Y., ANDREWS, C. A., MCLEOD, G., BORKOWSKY, W., FARTHING, C. & HO, D. D. 1994. Temporal association of cellular immune responses with the initial control of viremia in primary human immunodeficiency virus type 1 syndrome. *J Virol,* 68, 4650-5.

KOWALSKI, M., BERGERON, L., DORFMAN, T., HASELTINE, W. & SODROSKI, J. 1991. Attenuation of human immunodeficiency virus type 1 cytopathic effect by a mutation affecting the transmembrane envelope glycoprotein. *J Virol,* 65, 281-91.

KWONG, P. D., WYATT, R., ROBINSON, J., SWEET, R. W., SODROSKI, J. & HENDRICKSON, W. A. 1998. Structure of an HIV gp120 envelope glycoprotein in complex with the CD4 receptor and a neutralizing human antibody. *Nature,* 393, 648-59.

LASKY, L. A., GROOPMAN, J. E., FENNIE, C. W., BENZ, P. M., CAPON, D. J., DOWBENKO, D. J., NAKAMURA, G. R., NUNES, W. M., RENZ, M. E. & BERMAN, P. W. 1986. Neutralization of the AIDS retrovirus by antibodies to a recombinant envelope glycoprotein. *Science,* 233, 209-12.

LEONARD, C. K., SPELLMAN, M. W., RIDDLE, L., HARRIS, R. J., THOMAS, J. N. & GREGORY, T. J. 1990. Assignment of intrachain disulfide bonds and characterization of potential glycosylation sites of the type 1 recombinant human immunodeficiency virus envelope glycoprotein (gp120) expressed in Chinese hamster ovary cells. *J Biol Chem,* 265, 10373-82.

LEVESQUE, M. C., MOODY, M. A., HWANG, K. K., MARSHALL, D. J., WHITESIDES, J. F., AMOS, J. D., GURLEY, T. C., ALLGOOD, S., HAYNES, B. B., VANDERGRIFT, N. A., PLONK, S., PARKER, D. C., COHEN, M. S., TOMARAS, G. D., GOEPFERT, P. A., SHAW, G. M.,

SCHMITZ, J. E., ERON, J. J., SHAHEEN, N. J., HICKS, C. B., LIAO, H. X., MARKOWITZ, M., KELSOE, G., MARGOLIS, D. M. & HAYNES, B. F. 2009. Polyclonal B cell differentiation and loss of gastrointestinal tract germinal centers in the earliest stages of HIV-1 infection. *PLoS Med,* 6, e1000107.

LIU, S., FAN, S. & SUN, Z. 2003. Structural and functional characterization of the human CCR5 receptor in complex with HIV gp120 envelope glycoprotein and CD4 receptor by molecular modeling studies. *J Mol Model,* 9, 329-36.

LY, T. D., LAPERCHE, S. & COUROUCE, A. M. 2001. Early detection of human immunodeficiency virus infection using third- and fourth-generation screening assays. *Eur J Clin Microbiol Infect Dis,* 20, 104-10.

MALIM, M. H. & BIENIASZ, P. D. 2012. HIV Restriction Factors and Mechanisms of Evasion. *Cold Spring Harb Perspect Med,* 2, a006940.

MAO, Y., WANG, L., GU, C., HERSCHHORN, A., XIANG, S. H., HAIM, H., YANG, X. & SODROSKI, J. 2012. Subunit organization of the membrane-bound HIV-1 envelope glycoprotein trimer. *Nat Struct Mol Biol,* 19, 893-9.

MASUR, H., MICHELIS, M. A., GREENE, J. B., ONORATO, I., STOUWE, R. A., HOLZMAN, R. S., WORMSER, G., BRETTMAN, L., LANGE, M., MURRAY, H. W. & CUNNINGHAM-RUNDLES, S. 1981. An outbreak of community-acquired Pneumocystis carinii pneumonia: initial manifestation of cellular immune dysfunction. *N Engl J Med,* 305, 1431-8.

MCELRATH, M. J., DE ROSA, S. C., MOODIE, Z., DUBEY, S., KIERSTEAD, L., JANES, H., DEFAWE, O. D., CARTER, D. K., HURAL, J., AKONDY, R., BUCHBINDER, S. P., ROBERTSON, M. N., MEHROTRA, D. V., SELF, S. G., COREY, L., SHIVER, J. W., CASIMIRO, D. R. & STEP STUDY PROTOCOL, T. 2008. HIV-1 vaccine-induced immunity in the test-of-concept Step Study: a case-cohort analysis. *Lancet,* 372, 1894-905.

MODROW, S., HAHN, B. H., SHAW, G. M., GALLO, R. C., WONG-STAAL, F. & WOLF, H. 1987. Computer-assisted analysis of envelope protein sequences of seven human immunodeficiency virus isolates: prediction of antigenic epitopes in conserved and variable regions. *J Virol,* 61, 570-8.

MOIR, S., MALASPINA, A., LI, Y., CHUN, T. W., LOWE, T., ADELSBERGER, J., BASELER, M., EHLER, L. A., LIU, S., DAVEY, R. T., JR., MICAN, J. A. & FAUCI, A. S. 2000. B cells of HIV-1-infected patients bind virions through CD21-complement interactions and transmit infectious virus to activated T cells. *J Exp Med,* 192, 637-46.

MOORE, J. P., SATTENTAU, Q. J., WYATT, R. & SODROSKI, J. 1994. Probing the structure of the human immunodeficiency virus surface glycoprotein gp120 with a panel of monoclonal antibodies. *J Virol,* 68, 469-84.

MORIKITA, T., MAEDA, Y., FUJII, S., MATSUSHITA, S., OBARU, K. & TAKATSUKI, K. 1997. The V1/V2 region of human immunodeficiency virus type 1 modulates the sensitivity to neutralization by soluble CD4 and cellular tropism. *AIDS Res Hum Retroviruses,* 13, 1291-9.

MOSER, B. 1998. Human chemokines: role in lymphocyte trafficking. *Sci Prog,* 81 (Pt 4), 299-313.

NEIL, S. J., SANDRIN, V., SUNDQUIST, W. I. & BIENIASZ, P. D. 2007. An interferon-alpha-induced tethering mechanism inhibits HIV-1 and Ebola virus particle release but is counteracted by the HIV-1 Vpu protein. *Cell Host Microbe,* 2, 193-203.

OKOYE, A. A. & PICKER, L. J. 2013. CD4(+) T-cell depletion in HIV infection: mechanisms of immunological failure. *Immunol Rev,* 254, 54-64.

POLES, M. A., ELLIOTT, J., TAING, P., ANTON, P. A. & CHEN, I. S. 2001. A preponderance of CCR5(+) CXCR4(+) mononuclear cells enhances gastrointestinal mucosal susceptibility to human immunodeficiency virus type 1 infection. *J Virol,* 75, 8390-9.

REITTER, J. N., MEANS, R. E. & DESROSIERS, R. C. 1998. A role for carbohydrates in immune evasion in AIDS. *Nat Med,* 4, 679-84.

RERKS-NGARM, S., PITISUTTITHUM, P., NITAYAPHAN, S., KAEWKUNGWAL, J., CHIU, J., PARIS, R., PREMSRI, N., NAMWAT, C., DE SOUZA, M., ADAMS, E., BENENSON, M., GURUNATHAN, S., TARTAGLIA, J., MCNEIL, J. G., FRANCIS, D. P., STABLEIN, D., BIRX, D. L., CHUNSUTTIWAT, S., KHAMBOONRUANG, C., THONGCHAROEN, P., ROBB, M. L., MICHAEL, N. L., KUNASOL, P., KIM, J. H. & INVESTIGATORS, M.-T. 2009. Vaccination with ALVAC and AIDSVAX to prevent HIV-1 infection in Thailand. *N Engl J Med,* 361, 2209-20.

ROBERT GRAHAM QUINTON LESLIE. 2001. Complement Receptors. *ENCYCLOPEDIA OF LIFE SCIENCES* [Online].

ROBEY, W. G., SAFAI, B., OROSZLAN, S., ARTHUR, L. O., GONDA, M. A., GALLO, R. C. & FISCHINGER, P. J. 1985. Characterization of envelope and core structural gene products of HTLV-III with sera from AIDS patients. *Science,* 228, 593-5.

SATTENTAU, Q. J., DALGLEISH, A. G., WEISS, R. A. & BEVERLEY, P. C. 1986. Epitopes of the CD4 antigen and HIV infection. *Science,* 234, 1120-3.

SCHMITZ, J. E., KURODA, M. J., SANTRA, S., SASSEVILLE, V. G., SIMON, M. A., LIFTON, M. A., RACZ, P., TENNER-RACZ, K., DALESANDRO, M., SCALLON, B. J., GHRAYEB, J., FORMAN, M. A., MONTEFIORI, D. C., RIEBER, E. P., LETVIN, N. L. & REIMANN, K. A. 1999. Control of viremia in simian immunodeficiency virus infection by CD8+ lymphocytes. *Science,* 283, 857-60.

SCHREIBER, M., MULLER, H., WACHSMUTH, C., LAUE, T., HUFERT, F. T., VAN LAER, M. D. & SCHMITZ, H. 1997. Escape of HIV-1 is associated with lack of V3 domain-specific antibodies *in vivo. Clin Exp Immunol,* 107, 15-20.

SICKINGER, E., JONAS, G., YEM, A. W., GOLLER, A., STIELER, M., BRENNAN, C., HAUSMANN, M., SCHOCHETMAN, G., DEVARE, S. G., HUNT, J. C., KAPPRELL, H. P. & BRYANT, J. D. 2008. Performance evaluation of the new fully automated human immunodeficiency virus antigen-antibody combination assay designed for blood screening. *Transfusion,* 48, 584-93.

SMITH, B. A., GARTNER, S., LIU, Y., PERELSON, A. S., STILIANAKIS, N. I., KEELE, B. F., KERKERING, T. M., FERREIRA-GONZALEZ, A., SZAKAL, A. K., TEW, J. G. & BURTON, G. F. 2001. Persistence of infectious HIV on follicular dendritic cells. *J Immunol,* 166, 690-6.

SPECK, R. F., WEHRLY, K., PLATT, E. J., ATCHISON, R. E., CHARO, I. F., KABAT, D., CHESEBRO, B. & GOLDSMITH, M. A. 1997. Selective employment of chemokine receptors as human immunodeficiency virus type 1 coreceptors determined by individual amino acids within the envelope V3 loop. *J Virol,* 71, 7136-9.

STEVCEVA, L., YOON, V., ANASTASIADES, D. & POZNANSKY, M. C. 2007. Immune responses to HIV Gp120 that facilitate viral escape. *Curr HIV Res,* 5, 47-54.

STEVCEVA, L., YOON, V., CARVILLE, A., PACHECO, B., SANTOSUOSSO, M., KORIOTH-SCHMITZ, B., MANSFIELD, K. & POZNANSKY, M. C. 2008. The efficacy of T cell-mediated immune responses is reduced by the envelope protein of the chimeric HIV-1/SIV-KB9 virus *in vivo. J Immunol,* 181, 5510-21.

SUNDQUIST, W. I. & KRAUSSLICH, H. G. 2012. HIV-1 assembly, budding, and maturation. *Cold Spring Harb Perspect Med,* 2, a006924.

TAKEMURA, T. & MURAKAMI, T. 2012. Functional constraints on HIV-1 capsid: their impacts on the viral immune escape potency. *Front Microbiol,* 3, 369.

TERAI, C., KORNBLUTH, R. S., PAUZA, C. D., RICHMAN, D. D. & CARSON, D. A. 1991. Apoptosis as a mechanism of cell death in cultured T lymphoblasts acutely infected with HIV-1. *J Clin Invest,* 87, 1710-5.

TURVILLE, S. G., CAMERON, P. U., HANDLEY, A., LIN, G., POHLMANN, S., DOMS, R. W. & CUNNINGHAM, A. L. 2002. Diversity of receptors binding HIV on dendritic cell subsets. *Nat Immunol,* 3, 975-83.

WALKER, B. D. 2007. Elite control of HIV Infection: implications for vaccines and treatment. *Top HIV Med,* 15, 134-6.

WAWER, M. J., GRAY, R. H., SEWANKAMBO, N. K., SERWADDA, D., LI, X., LAEYENDECKER, O., KIWANUKA, N., KIGOZI, G., KIDDUGAVU, M., LUTALO, T., NALUGODA, F., WABWIRE-MANGEN, F., MEEHAN, M. P. & QUINN, T. C. 2005. Rates of HIV-1 transmission per coital act, by stage of HIV-1 infection, in Rakai, Uganda. *J Infect Dis,* 191, 1403-9.

WEI, X., DECKER, J. M., WANG, S., HUI, H., KAPPES, J. C., WU, X., SALAZAR-GONZALEZ, J. F., SALAZAR, M. G., KILBY, J. M., SAAG, M. S., KOMAROVA, N. L., NOWAK, M. A., HAHN, B. H., KWONG, P. D. & SHAW, G. M. 2003. Antibody neutralization and escape by HIV-1. *Nature,* 422, 307-12.

WHO 2015. HIV/AIDS Factsheet No 360. *In* WHO (ed.) Geneva Switzerland: WHO.

WILLEY, S., AASA-CHAPMAN, M. M., O'FARRELL, S., PELLEGRINO, P., WILLIAMS, I., WEISS, R.
    A. & NEIL, S. J. 2011. Extensive complement-dependent enhancement of HIV-1 by autologous
    non-neutralising antibodies at early stages of infection. *Retrovirology,* 8, 16.
WRIGHT, E. R., SCHOOLER, J. B., DING, H. J., KIEFFER, C., FILLMORE, C., SUNDQUIST, W. I. &
    JENSEN, G. J. 2007. Electron cryotomography of immature HIV-1 virions reveals the structure of
    the CA and SP1 Gag shells. *EMBO J,* 26, 2218-26.
WU, L., GERARD, N. P., WYATT, R., CHOE, H., PAROLIN, C., RUFFING, N., BORSETTI, A.,
    CARDOSO, A. A., DESJARDIN, E., NEWMAN, W., GERARD, C. & SODROSKI, J. 1996. CD4-
    induced interaction of primary HIV-1 gp120 glycoproteins with the chemokine receptor CCR-5.
    Nature, 384, 179-83.
WYATT, R., KWONG, P. D., DESJARDINS, E., SWEET, R. W., ROBINSON, J., HENDRICKSON, W. A.
    & SODROSKI, J. G. 1998. The antigenic structure of the HIV gp120 envelope glycoprotein.
    Nature, 393, 705-11.
WYATT, R. & SODROSKI, J. 1998. The HIV-1 envelope glycoproteins: fusogens, antigens, and
    immunogens. *Science,* 280, 1884-8.
WYATT, R., SULLIVAN, N., THALI, M., REPKE, H., HO, D., ROBINSON, J., POSNER, M. &
    SODROSKI, J. 1993. Functional and immunologic characterization of human immunodeficiency
    virus type 1 envelope glycoproteins containing deletions of the major variable regions. *J Virol,* 67,
    4557-65.
YANG, Z. Y., CHAKRABARTI, B. K., XU, L., WELCHER, B., KONG, W. P., LEUNG, K., PANET, A.,
    MASCOLA, J. R. & NABEL, G. J. 2004. Selective modification of variable loops alters tropism
    and enhances immunogenicity of human immunodeficiency virus type 1 envelope. *J Virol,* 78,
    4029-36.
YUAN, T., LI, J. & ZHANG, M. Y. 2013. HIV-1 envelope glycoprotein variable loops are indispensable for
    envelope structural integrity and virus entry. *PLoS One,* 8, e69789.
ZHANG, L., HUANG, Y., HE, T., CAO, Y. & HO, D. D. 1996. HIV-1 subtype and second-receptor use.
    Nature, 383, 768.
ZHU, P., LIU, J., BESS, J., JR., CHERTOVA, E., LIFSON, J. D., GRISE, H., OFEK, G. A., TAYLOR, K.
    A. & ROUX, K. H. 2006. Distribution and three-dimensional structure of AIDS virus envelope
    spikes. *Nature,* 441, 847-52.

CHAPTER 8

Vaccines Against Latent Viral Infections: Pathway Forward

Liljana Stevceva[*]

California Northstate University College of Medicine, Elk Grove, CA, USA

Abstract: With new evidence emerging that viral latent infections might have yet unknown negative effects, it is becoming a necessity to look into the factors that allow these viruses to persist and establish latency. Designing vaccines that will prevent the initial infection and establishment of latency should become the main focus. To succeed in such a goal is not impossible but requires major refocusing of the research efforts. The viral escape mechanisms that are used during the initial stages of the acute primary infection should be carefully examined and fully understood. Once this is accomplished, developing strategies that will disarm these escape mechanisms and allow the immune system to clear the virus should become an achievable goal.

Keywords: EBV, VZV, HIV, CMV, HSV-1, HSV-2, viral latency, persistent viral infection, chronic viral infection, reactivation, mucosa, herpesviridae, herpes zoster, varicella, Varivax®, Zostavax®, retroviruses, gp350, gp220, gp85 (gH), gp25 (gL), gp42/38, CR2, CD21, C3, C3d, C3b, C3a, C3c, iC3b, C5a, C3dg, C3d, gB, gC, gD, gE, gG, gH, gI, gJ, gK, gL, gM, gH-gL, gM-gN, heparan sulphate, lectin pathway, IgA, IgG Fc, DC-SIGN, Trojan horse, CCR5, CXCR4, TGFβ, regulatory T cells, virokines, viroceptors, CD103+ DCs, DAF, MCP, amplification loop, complement

VIRUS ESCAPE FROM THE IMMUNE RESPONSE AND ESTABLISHMENT OF LATENCY

Viral latency is the ability of the virus to remain dormant in a cell. During latency the virus does not replicate but it is also not eradicated meaning that it can reactivate, begin to multiply and cause a disease if favorable conditions are created. The data reviewed in the previous chapters clearly point to similarities in mechanisms by which the viruses such as EBV, CMV, HSV, VZV and HIV escape the immune response launched against them. Most of the viruses capable of establishing latent infections in humans belong to *Herpesviridae* family. Such are EBV, CMV, HSV and VZV. *Retroviruses* are also known to establish latent

*Corresponding author Liljana Stevceva: University of Texas Rio Grande Valley School of Medicine, 2102 Treasure Hills Blvd., Harlingen TX, USA; E-mail: liljana@hotmail.com

infections. The best known of them is HIV but humans have lived in a successful symbiosis with retroviruses since the beginning of times and today about 8% of the human genome consists of retroviral inserts that managed to get incorporated in it during the human evolution (Griffiths, 2001).

Until recently viruses causing latent infections in humans were considered harmless and this is why there were no serious attempts to develop strategies to prevent or treat the viral latency. As reviewed in Chapter 1 of the book recent emerging evidence on the effects of CMV latency on the immune system shattered that belief and inspired further research into the effects of latent viruses on the immune responses including those to vaccines (Looney *et al.*, 1999), (Derhovanessian *et al.*, 2013), (Frasca *et al.*, 2015).

In this book we attempted to examine the important structural characteristics of HSV, CMV, EBV, HIV and VZV and to look for common structures that function in favor of viral escape from immune recognition and destruction.

To start with and as summarized in Chapter 2, all of these viruses are enveloped viruses. The envelope always contains glycoprotein spikes. Newly emerging research is indicating that the glycoprotein spikes on the viral envelopes of these viruses might be interfering with the complement-mediated killing of viruses and virus-infected host cells. The complement cascade is one of the oldest defense mechanisms of the human body and it is believed to have evolved 700 million years ago. It is part of both innate immune mechanisms (alternative and lectin pathway of activation) and adaptive immunity (classical pathway). In essence, the complement system consists of about 30 enzymes that become activated sequentially on the surfaces of pathogens or the infected host cells and lead to destruction of the pathogen/the infected cells by two main mechanisms:

1. Complement labels pathogens/host cells with C3b that binds on their surface (opsonization) and this way they become visible to the immune system for destruction (phagocytosis; antibodies can bind to them)

2. Activation of the complement component 5 (C5) by C5 convertase starts the formation of the membrane attack system that creates a hole in the membrane of the pathogen/host cell thus allowing the water from the intracellular space to enter into the pathogen/host cell and to destroy them by osmolysis

There are three main pathways by which the complement cascade becomes activated. The classical pathway is initiated by activation of the complement component 1 (C1q) usually when C1q binds to IgG or IgM complex with antigens or if C1q directly binds to the surface of the pathogen.

The alternative pathway of complement activation occurs continuously at a low level and leads to deposition of C3b on pathogen/host cell surfaces. This activation is consistently and carefully down regulated. The failure of the regulatory mechanisms of the activation of alternative pathway will lead to uncontrolled complement cascade resulting in pathogen/host cell or tissue destruction.

The lectin pathway of complement activation occurs when mannose-binding lectin (MBL) binds to mannose residues on the surface of pathogens.

All three pathways of complement activation result in the formation of C3 convertase as a central step. C3 convertase cleaves the complement component 3 (C3) to C3a and C3b and the newly created C3b binds to the surface of the pathogen/host cell 'opsonizing' it. This also starts the amplification loop of the complement cascade (see Fig. **1**) that, under normal circumstances functions in such a way to allow buildup of sufficient amount of the opsonin C3b so that the pathogen/host cell can get 'noticed' by the immune system.

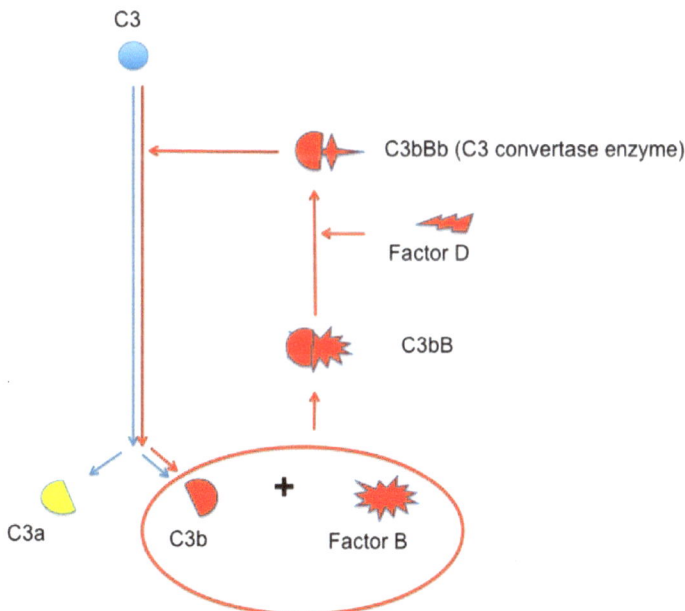

Fig. (1). The amplification loop of the complement cascade. The purpose of this loop under normal circumstances is to generate enough C3b so that C5 convertase can be formed (excess C3b is needed for that to occur) and the creation of the membrane attack complex can be started. In order for that to happen the C3b binds to Factor B to form C3bB. Factor D catalyses another reaction leading to the formation of C3bBb, the C3 convertase enzyme that cleaves more C3 into C3a and C3b. If the loop is not down regulated at one point, it will lead to consumption of all available C3.

As illustrated in Figs. (**2a**) and (**2b**) the amplification loop is normally heavily regulated and as soon as enough C3b is generated to create the C5 convertase, the loop is down regulated.

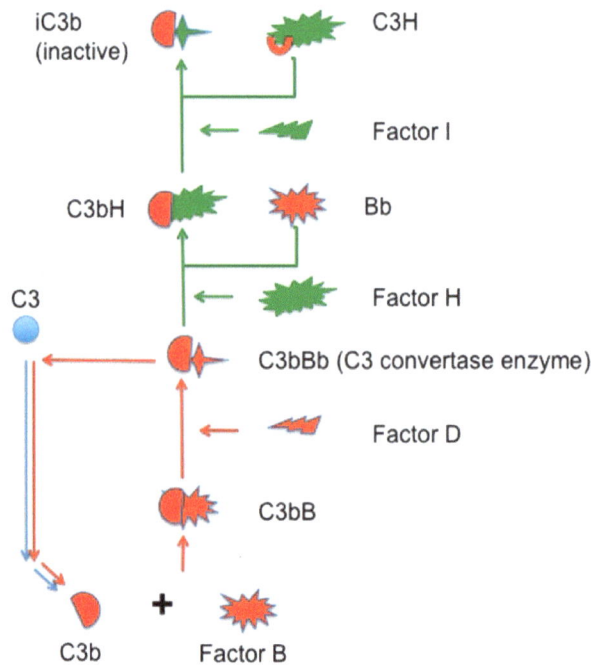

Fig. (2a). Regulation of the amplification loop by soluble factors. The amplification loop is regulated by Factor H that binds to C3 convertase and pushes out Factor B, forming C3bH. Another regulating soluble enzyme is Factor I that binds to C3bH and inactivates C3b into its inactive form iC3b while also creating C3H.

Creation of the C5 convertase is the first step in the complement cascade towards generation of the membrane attack complex. Once formed, the C3 convertase remains attached to the pathogen/host cell membrane. This complex might interact with another C3b and thus form the alternative pathway C5 convertase. The C5 convertase cleaves complement component 5 (C5) and this is the first step to generation of the membrane attack complex (see Fig. **3**). As previously mentioned, successful insertion of the membrane attack complex leads to lysis of the pathogen/host cell.

Fig. (2b). Regulation of the amplification loop by non-soluble factors. Membrane receptor CD55 (Decay Accelerating Factor; DAF) disrupts the interaction of C3b and Factor B. The membrane receptor CD46 (MCP) is a co-factor for Factor I that plays a role in cleavage of C3b and C4b to their inactive form, iC3b, iC4b. The membrane receptor CD35 (CR1) accelerates the decay of C3 convertase and acts as a cofactor for Factor I.

Over the millions of years of co-existence of viruses and humans, viruses have developed some fairly sophisticated mechanisms to escape the complement destruction (Favoreel *et al.*, 2003), (Blue *et al.*, 2004). Although much more research into this is needed, this appears to be especially true for the viruses that have the ability to go into latency. There is evidence that all of the viruses reviewed in this book utilize the glycoprotein spikes on their surface to somehow escape or skew the complement cascade.

In **EBV** the most dominant glycoprotein spikes are gp350 and gp220. Other, minor glycoprotein spikes are gp85 (gH), gp25 (gL) and gp42/38. The gp350/220 glycoprotein binds to the C3d complement receptor CR2 (CD21) on the B cells (Nemerow *et al.*, 1985). This binding does not result in efficient infection of the host cell so, it is believed that its specific role is just to attach the virion to the B cell surface (Speck *et al.*, 2000). Entry of EBV into the host cell requires interaction of three other glycoproteins: gH, gL and gp42. The glycoprotein gH fuses with gL and this complex is responsible for viral membrane fusion to B cells. Then, the gH-gL complex interacts with gp42 and this interaction is required for penetration of the virus into the host B cell as reviewed in (Speck *et al.*, 2000).

Entry into epithelial cells is unclear with some studies demonstrating a role for gH into the entry process and others reporting that epithelial cells expressing the IgA receptor may internalize EBV bound on IgA (Sixbey and Yao, 1992). Cell to cell infection with EBV has also being demonstrated and requires direct cell-to-cell contact. The exact mechanism for this is not clear. The complement component C3d is a byproduct of cleavage of the C3, a complement component that is central in the complement cascade and requires activation for the classical and alternative complement activation. During both of these pathways, C3 is cleaved to C3a and C3b components that are also required for the lectin pathway. Of these C3b is opsonin and facilitates phagocytosis. C3b is broken down progressively to first iC3b, then C3c + C3dg, and then finally C3d. C3d binds to the CR2 receptor on B cells and is believed to facilitate phagocytosis by B cells. Because of the importance of the C3d-CR2 interaction in improving the immune response to antigens, this bond is often targeted by pathogens. CR2 also binds iC3b and C3dg complement components. Binding of all three components to CR2 lowers the threshold of antibody stimulation and the amount of antigen required to activate the B cells 10-100 fold. In addition, binding of CR2 receptor rescues B cells from apoptosis. The EBV gp350/220 glycoprotein has several amino acids homology to the C3dg fragment and this is why it readily binds to CR2. Binding of C3dg photolytic fragment induces proliferation of B cells. The EBV-induced B cell proliferation causes polyclonal B cell expansion and production of antibodies. In addition, binding of EBV gp350/220 glycoprotein to CR2 occupies the receptor that in conjunction with CR1 plays a significant role in limiting the amplification loop of the complement cascade (for the role of CR1 see Fig. **2b**).

HSV is another potent latent virus that disrupts the complement cascade by interfering with the C3 components. HSV envelope has 600-750 glycoprotein spikes on its surface consisting of at least 11 different glycoproteins such as gB, gC, gD, gE, gG, gH, gI, gJ, gK, gL and gM. Two glycoproteins bind to heparan sulphate and mediate binding of the virion to the cell surface (Spear, 2004). The glycoprotein gB also plays a role in viral entry into the host cell. Interaction of gD and the gH-gL complex are also required for fusion of the viral envelope with the host cell membrane. All these glycoproteins are highly conserved: gH, gL and gB in all *Herpesviridae* while gC and gD in all *Alphaherpesviridae*.

Especially gC and gE have been shown to play an important role in viral escape from complement-mediated killing. During HSV infection, complement cascade gets activated in presence or absence of antibody. This leads to cleavage of C3 into C3a and C3b as described above. C3b remains bound to the activating surface. The virion gC binds to C3b and interrupts the complement cascade by inhibiting the creation of the C5 convertase, which is the first step to generating the membrane attack complex that could potentially lyse the virus (see Fig. **3**). The gC also has the capacity to bind to C5 and this contributes to the interruption of the complement cascade (Friedman, 2003). Antibody-mediated activation of the complement cascade is also interrupted by binding of gE to IgG Fc domain.

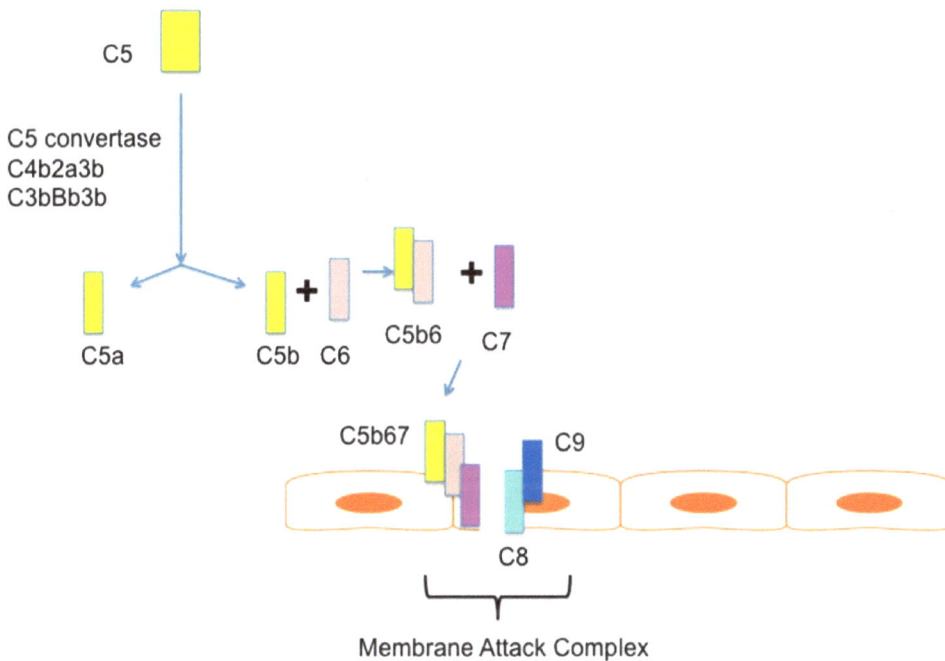

Fig. (3). Formation of the membrane attack complex. The C5 convertase (C4b2a3b or C3bBb3b) cleaves C5 into C5a and C5b. The complement component C5b then associates with C6 forming C5b6 and with C7 creating C5b67 that inserts itself into the pathogen/host cell membrane. Activation of C8 and C9 completes this process creating a hole in the membrane. This allows fluid from the intracellular space to enter the pathogen/host cell and to destroy it by osmolysis.

HCMV has more than 50 glycoproteins in its envelope. Of these gM is the most abundant and makes up 10% of the mass of the virus. It has been shown that gM is essential for growth of HCMV in fibroblasts. In addition, gM makes a complex with gN and this complex binds to heparan sulphate on the host cell membrane. Thus, the gM-gN complex plays a role in the initial attachment of the virion to the host cell. Other than these two roles of this, most abundant glycoprotein on the surface of HCMV, not much is known about gM and its function. As with other members of *Herpesviridae* gH and gL glycoproteins play a role in fusion of the viral envelope with the host cell membrane. Because primary infection with HCMV is usually asymptomatic and the virus only reactivates when a condition of immunodeficiency is established, not much research effort has been devoted to investigating the mechanisms of viral escape during establishment of latency. It has been reported that the first response to primary HCMV infection consists of virus-specific CD4+ T cells that appear about 10 days after HCMV DNA is first detected. This is followed by IgM and IgG antibody response that appears about 7 days after the appearance of the CD4+ T cells and CD8+ T cell response 14 days after the CD4+ T cells response (Gamadia *et al.*, 2003). Most of the antibodies during the primary immune response to HCMV are directed towards the gB glycoprotein. The glycoprotein gB ligand for the C-type lectin dendritic cell-specific intercellular adhesion molecule-3-grabbing non-integrin (DC-SIGN). DC-SIGN has a prominent role in the docking of HCMV on monocyte-derived DCs (MDDCs). Furthermore, the virions enter into dendritic cells through micropinocytosis-like mechanism and remain intact and infectious for several hours. Such infectious particles can be released and are able to trans-infect other cells such as fibroblasts (Haspot *et al.*, 2012). Thus, HCMV uses DC-SIGN (similarly to HIV) as a 'Trojan Horse' to spread the infection. Other HCMV glycoproteins US2, US3, US6 and US11 down modulate the cell surface expression of MHC Class 1 proteins altering T cell priming, CTL killing and NK cell activity. In addition, the virus uses a whole battery of weapons to escape the immune responses and to establish latency such as virokines (cytokine homologs), soluble cytokine scavengers that seize free chemokines and viroceptors (homologs of cell surface receptors) (Lucas *et al.*, 2001). HCMV activates the classical pathway of complement as early as 4 hours after infection by direct binding of the C1q (Spiller and Morgan, 1998). In addition, in mice, CMV infection upregulates expression of the complement regulatory proteins DAF (CD55) and MCP (CD46) that protect the host cell from complement mediated lysis. DAF inhibits the creation of the C5 convertase by blocking the association of C3b to Factor B and creation of C3 convertase and accelerates dissociation of Bb from the already formed C3 convertase. C3 convertase and surplus C3b are required for formation of the C5 convertase C3bBb3b that dissociates C5 in the first step in the formation of the membrane attack complex. MCP is a cofactor to Factor I that inactivates C3b and C4b thus preventing formation of both types of C5 convertase, C3bBb3b and C4b2a3b (illustrated in Fig. **2a** and **2b**).

VZV envelope contains several glycoproteins such as gB, gC, gE, gH, gI, gK, gL, gM and gN. Of these, gE is the most abundant and similarly to the one present on the HSV envelope, it binds to the Fc portion of IgG. Whether its function is to prevent antibody-mediated complement killing as in HSV remains to be investigated. gI is required for maturation and efficient distribution of gE to the cell surface. As described in Chapter 6, gI is essential for cell-to-cell spreading of the virus, adsorption to cells and syncytia formation *in vitro*. gI and gE form a complex and this complex. VZV gB is the second most abundant glycoprotein and it binds to heparan sulphate facilitating attachment of the virion to the host cell. The gH/gL complex is required for fusion of the viral envelope to the host cell membrane. The glycoprotein gC is binding to mannose-6 phosphate receptor (MPR) and it has not been investigated in depth. It has been reported that gC plays a role in infectivity of the virus to the skin (Moffat *et al.*, 1998).

HIV has about 14 glycoprotein spikes (gp120) on its surface that are anchored to the membrane *via* gp41 protein. The envelope glycoprotein of HIV plays an important role in viral escape and actively shields the virus from immune attack. As discussed in details in Chapter 6, gp120 is essential for the infectivity of the virus and the gp120 binding to the CD4 receptor and a co-receptor (CCR5 or CXCR4) is required for fusion of viral envelope to the host cell membrane. The glycoprotein induces polyclonal B cell activation and it activates the complement cascade resulting in deposition of complement components to the viral surface and creation of complement-antibody complexes that do not really kill the virus because the initial antibodies are not neutralizing but are actually facilitating binding of the virus to cells expressing complement receptors C3 and C4. Similarly to HCMV, binding of HIV gp120 to DC-SIGN facilitates phagocytosis of the still infectious virus and its spread to CD4+ T cells *via* the normal antigen presentation process. Similarly to EBV, the complement opsonized HIV binds to the CD21 receptor on B cells and is carried on their surface until it infects T cells trough the normal B-T cell interaction.

Thus, the HIV uses its glycoprotein spikes to hijack the main players of the immune response (DCs, B cells) and utilizes the normal cell interactions and immune mechanisms not only to escape the immune response but also to spread.

As reviewed above, the envelope glycoproteins of viruses that are able to cause latent and persistent infections seem to have an essential role in enabling viral infection and in many cases in surpassing the immune response or even in utilizing it to spread and survive. Many of those glycoproteins are highly conserved for that reason and evolutionary this probably means that the virus cannot be harmed by the immune response directed towards those conserved epitopes. Thus, the popular approach to use the conserved regions of the envelope to create successful vaccine might be fundamentally flawed.

Much more research needs to be directed to explore the common mechanisms that these viruses use to evade the immune system.

CONTRIBUTING FACTORS OF THE MUCOSAL ENTRY SITE TO ESTABLISHMENT OF LATENCY

All viruses capable of establishing a latent and persistent infection enter the human body through mucosae. Although it is quite clear that mucosal entry site is not a pre requisite for establishing latency, it would be interesting to explore the specificities of the mucosal immune milieu that contribute to the viral escape during the acute stages of viral infection. Mucosal surfaces are continuously exposed to antigens from food or from pathogens especially oropharyngeal mucosa, intestinal mucosa, vaginal and cervical mucosa. Because of that, mucosae are constantly in a state of low inflammation and the mucosal immune response is often directed to tolerating the commensal flora. This is mediated through dendritic cells that can extend their processes into the lumen in order to sample the antigens (Rescigno *et al*., 2001), (Niess *et al*., 2005). DC-SIGN is expressed on dendritic cells on all mucosal surfaces and lymphoid tissues but is absent on most dendritic cells in peripheral blood except for DC2 (Soilleux *et al*., 2002). Thus, viruses such as HIV and CMV could bind to DC-SIGN and use the dendritic cell as a vehicle to enter the lymphoid tissues where they gain access to T cells *via* the antigen presentation process. In the gut mucosa, the dendritic cells located in lamina propria express CD103 marker and do not travel beyond mesenteric lymph nodes. These DCs are particularly potent in generating FoxP3+ regulatory T cells that promote tolerance as reviewed in (Pabst and Mowat, 2012). Similarly, vaginal DCs rapidly bind HIV-1 and transmit it to vaginal and peripheral blood lymphocytes (Shen *et al*., 2014). DCs expressing CD103 also exist in buccal and gingival mucosa. Also, expression of Toll-like receptors 2 and 4 (TLR2 and TLR4) is higher on oral Langerhans cells. TLR triggering with lipopolysaccharide upregulates expression of the co-inhibitory molecules B7-H1 and B7-H3, whereas expression of the co-stimulatory molecule CD86 (B7-2) is decreased resulting in development of Tregs and secretion of interleukin IL-10 and TGFβ by stimulated T cells (Hovav, 2014).

In addition, we have shown in our studies that HIV-1 gp120 glycoprotein directly suppresses T cell secretion of IFNγ, TNFα and IL-10 and causes failure of T cells to degranulate in a dose dependent manner. Gp120 also induces TGFβ secretion and subsequently increases the numbers of regulatory T cells and tolerance. This seems to be especially prominent in the lymphoid tissue during acute SHIV infection (Stevceva *et al.*, 2008).

All of the factors outlined above that are specific for the mucosal environment might be contributing to tolerating the invading viruses. These, in addition to their specific structural properties, create a very permissive environment for these viruses to evade destruction by the immune system and to instead establish equilibrium and latency.

CURRENT VACCINES AGAINST VIRUSES THAT ESTABLISH LATENT INFECTIONS

As previously mentioned, research efforts to develop vaccines related to latent viral infections were so far focused on preventing the clinical manifestation of the disease. The most widely used, the most successful and the only approved vaccine so far has been the vaccine against varicella, Varivax®. The varicella vaccine consists of live, attenuated OKA/Merck varicella virus (Quinlivan *et al.*, 2011). It was approved for use in the United States in 1995 for persons 12 months or older that have not had varicella. Second dose was recommended in the USA in 2006. The vaccine is given subcutaneously, and it is recommended that the first dose be given to children 12-15 months old and the second dose to 4-6 years old. The two doses should be separated for at least 3 months. A 14 years prospective cohort study conducted at Kaiser Permanente, California looked at the effectiveness of Varivax® in preventing chicken pox. The study followed 7585 children vaccinated with Varivax® from 1995 till 2009. Of these children, 2826 received a second dose of the vaccine. The average rate of varicella in this population was eight to ten fold lower than what was previously reported. The vaccine efficacy was estimated at 90%. No varicella cases were reported in children that received both doses of the vaccine. The vaccinated children still developed herpes zoster. A total of 113 cases of herpes zoster were reported in the 14 year follow up. Although the authors claim reduction of 40% in comparison with historical data (about 75 cases expected), they are only calculating the confirmed herpes zoster cases to arrive to that number. While providing valuable information, the study has a few significant flaws such as using historical data for comparison (no control group), not taking into account herd immunity and short follow up after the second dose was introduced (4 years until the study was published as opposed to 14 years for one dose only) (Baxter *et al.*, 2013). A much higher dose (at least 19,400 plaque-forming units compared to 1350 plaque-forming units in varicella vaccine) of the same attenuated virus is given as a vaccine aimed at reducing the incidence of herpes zoster in the elderly (Zostavax®). The rationale behind using Zostavax® was to boost the waning immunity against VZV in the aging population and by doing so, to empower the immune system of the body to keep the latent VZV under control and to prevent reactivation. In a randomized, double-blind, placebo-controlled trial that enrolled 38,546 participants aged 60 years or older, the incidence of herpes zoster decreased by 51.3%, the incidence of postherpertic neuralgia decreased by 66.5% and the burden of illness due to herpes zoster decreased by 61.1% in the vaccinated group compared to the group that received placebo (Oxman *et al.*, 2005).

While both of these vaccines are, at least partially, successful in preventing the clinical manifestation of the VZV infection (varicella and herpes zoster), the virus (now replaced with an attenuated version) still establishes latent infection and remains dormant in the human body for the duration of the lifespan of the individual. The consequences of this are that, potentially, the virus could still reactivate in situations where the immunity is compromised and that we do not really know if the sole presence of the dormant VZV in the dorsal ganglia of the nervous system is sufficient for that to occur.

Despite long-term research efforts to develop preventative vaccines against EBV, HSV, CMV and HIV, no successful vaccines have been developed against the clinical manifestation of the disease caused by these viruses. No attempts have been made to develop vaccines that prevent the viral latency.

PREVENTING ESTABLISHMENT OF LATENCY

Humans have lived in symbiotic relationship with viruses for a very long time. For example, all retroviruses have the capacity to insert a DNA copy of themselves into the host genome. The host will then carry the inserted DNA copy as its own and will transfer it to its offsprings as a 'novel allele'. These DNA copies are called endogenous retroviruses and can persist for millions of years. They are also called 'fossil viruses' and are a footprint of previous exposure to retroviruses (Nelson *et al.*, 2004). Most of these endogenous retroviruses are inactivated through a process called recombination deletion and are no longer capable of producing infectious virus. In humans, remnants of endogenous retroviruses occupy about 8% of the human genome and constitute about 30 000 different retroviruses (20 human retroviruses families) embedded in each person's genome (Sverdlov, 2000), (Nelson *et al.*, 2004). Other viruses such as those reviewed in this book although not present in the human genome on a permanent basis are very successful in establishing a symbiosis with the host so that the virus remains dormant and is not inducing a significant disease. Or, at least, that was the widely held belief until recently. Recent investigations into CMV latency are revealing a potentially important role for this virus in immunosenescence and so far, it seems that this is just the tip of the iceberg. Because latent infections do not usually cause clinical symptoms (or at least, this is not apparent or readily associated with the presence of the virus) we know very little about the mechanisms that they use to escape the immune response and go dormant and the not so apparent effects on the human body. Also, since most of these viral infections are ubiquitous, we truly don't know what happens if we don't have them in our body. Are we going to be healthier, younger? Will we be able to delay immunosenescence and to considerably reduce the incidence of malignancy in aging individuals?

Numerous attempts have been made (some successful) to create vaccines against the clinical manifestations of disease or even against reactivation (Zostavax) but to date, there have not been attempts to design a vaccine that will prevent the establishment of the latent state of the virus. The varicella vaccine, although successful, does not prevent establishment of latent infection but simply replaces the more virulent VZV with the OKA strain.

It is easy to imagine that designing a successful vaccine against establishment of latent state in these viruses will require a very different approach, such that takes into account the escape mechanisms that help the virus to avoid clearance. Much greater effort will need to be devoted to fully understand the major escape mechanisms that the virus uses early in infection. With all that in mind however, and considering all of the failed efforts to develop a successful vaccine against HIV for example, this approach seems to be the best way forward.

CONFLICT OF INTEREST

The author confirms that this chapter contents have no conflict of interest.

ACKNOWLEDGEMENTS

Declared None.

REFERENCES

BAXTER, R., RAY, P., TRAN, T. N., BLACK, S., SHINEFIELD, H. R., COPLAN, P. M., LEWIS, E., FIREMAN, B. & SADDIER, P. 2013. Long-term effectiveness of varicella vaccine: a 14-Year, prospective cohort study. *Pediatrics,* 131, e1389-96.

BLUE, C. E., SPILLER, O. B. & BLACKBOURN, D. J. 2004. The relevance of complement to virus biology. *Virology,* 319, 176-84.

DERHOVANESSIAN, E., THEETEN, H., HAHNEL, K., VAN DAMME, P., COOLS, N. & PAWELEC, G. 2013. Cytomegalovirus-associated accumulation of late-differentiated CD4 T-cells correlates with poor humoral response to influenza vaccination. *Vaccine,* 31, 685-90.

FAVOREEL, H. W., VAN DE WALLE, G. R., NAUWYNCK, H. J. & PENSAERT, M. B. 2003. Virus complement evasion strategies. *J Gen Virol,* 84, 1-15.

FRASCA, D., DIAZ, A., ROMERO, M., LANDIN, A. M. & BLOMBERG, B. B. 2015. Cytomegalovirus (CMV) seropositivity decreases B cell responses to the influenza vaccine. *Vaccine.*

FRIEDMAN, H. M. 2003. Immune evasion by herpes simplex virus type 1, strategies for virus survival. *Trans Am Clin Climatol Assoc,* 114, 103-12.

GAMADIA, L. E., REMMERSWAAL, E. B., WEEL, J. F., BEMELMAN, F., VAN LIER, R. A. & TEN BERGE, I. J. 2003. Primary immune responses to human CMV: a critical role for IFN-gamma-producing CD4+ T cells in protection against CMV disease. *Blood,* 101, 2686-92.

GRIFFITHS, D. J. 2001. Endogenous retroviruses in the human genome sequence. *Genome Biol,* 2, REVIEWS1017.

HASPOT, F., LAVAULT, A., SINZGER, C., LAIB SAMPAIO, K., STIERHOF, Y. D., PILET, P., BRESSOLETTE-BODIN, C. & HALARY, F. 2012. Human cytomegalovirus entry into dendritic cells occurs *via* a macropinocytosis-like pathway in a pH-independent and cholesterol-dependent manner. *PLoS One,* 7, e34795.

HOVAV, A. H. 2014. Dendritic cells of the oral mucosa. *Mucosal Immunol,* 7, 27-37.

LOONEY, R. J., FALSEY, A., CAMPBELL, D., TORRES, A., KOLASSA, J., BROWER, C., MCCANN, R., MENEGUS, M., MCCORMICK, K., FRAMPTON, M., HALL, W. & ABRAHAM, G. N. 1999. Role of cytomegalovirus in the T cell changes seen in elderly individuals. *Clin Immunol,* 90, 213-9.

LUCAS, M., KARRER, U., LUCAS, A. & KLENERMAN, P. 2001. Viral escape mechanisms--escapology taught by viruses. *Int J Exp Pathol,* 82, 269-86.

MOFFAT, J. F., ZERBONI, L., KINCHINGTON, P. R., GROSE, C., KANESHIMA, H. & ARVIN, A. M. 1998. Attenuation of the vaccine Oka strain of varicella-zoster virus and role of glycoprotein C in alphaherpesvirus virulence demonstrated in the SCID-hu mouse. *J Virol,* 72, 965-74.

NELSON, P. N., HOOLEY, P., RODEN, D., DAVARI EJTEHADI, H., RYLANCE, P., WARREN, P., MARTIN, J., MURRAY, P. G. & MOLECULAR IMMUNOLOGY RESEARCH, G. 2004. Human endogenous retroviruses: transposable elements with potential? *Clin Exp Immunol,* 138, 1-9.

NEMEROW, G. R., WOLFERT, R., MCNAUGHTON, M. E. & COOPER, N. R. 1985. Identification and characterization of the Epstein-Barr virus receptor on human B lymphocytes and its relationship to the C3d complement receptor (CR2). *J Virol,* 55, 347-51.

NIESS, J. H., BRAND, S., GU, X., LANDSMAN, L., JUNG, S., MCCORMICK, B. A., VYAS, J. M., BOES, M., PLOEGH, H. L., FOX, J. G., LITTMAN, D. R. & REINECKER, H. C. 2005. CX3CR1-mediated dendritic cell access to the intestinal lumen and bacterial clearance. *Science,* 307, 254-8.

OXMAN, M. N., LEVIN, M. J., JOHNSON, G. R., SCHMADER, K. E., STRAUS, S. E., GELB, L. D., ARBEIT, R. D., SIMBERKOFF, M. S., GERSHON, A. A., DAVIS, L. E., WEINBERG, A., BOARDMAN, K. D., WILLIAMS, H. M., ZHANG, J. H., PEDUZZI, P. N., BEISEL, C. E., MORRISON, V. A., GUATELLI, J. C., BROOKS, P. A., KAUFFMAN, C. A., PACHUCKI, C. T., NEUZIL, K. M., BETTS, R. F., WRIGHT, P. F., GRIFFIN, M. R., BRUNELL, P., SOTO, N. E., MARQUES, A. R., KEAY, S. K., GOODMAN, R. P., COTTON, D. J., GNANN, J. W., JR., LOUTIT, J., HOLODNIY, M., KEITEL, W. A., CRAWFORD, G. E., YEH, S. S., LOBO, Z., TONEY, J. F., GREENBERG, R. N., KELLER, P. M., HARBECKE, R., HAYWARD, A. R., IRWIN, M. R., KYRIAKIDES, T. C., CHAN, C. Y., CHAN, I. S., WANG, W. W., ANNUNZIATO, P. W., SILBER, J. L. & SHINGLES PREVENTION STUDY, G. 2005. A vaccine to prevent herpes zoster and postherpetic neuralgia in older adults. *N Engl J Med,* 352, 2271-84.

PABST, O. & MOWAT, A. M. 2012. Oral tolerance to food protein. *Mucosal Immunol,* 5, 232-9.

QUINLIVAN, M., BREUER, J. & SCHMID, D. S. 2011. Molecular studies of the Oka varicella vaccine. *Expert Rev Vaccines,* 10, 1321-36.

RESCIGNO, M., URBANO, M., VALZASINA, B., FRANCOLINI, M., ROTTA, G., BONASIO, R., GRANUCCI, F., KRAEHENBUHL, J. P. & RICCIARDI-CASTAGNOLI, P. 2001. Dendritic cells express tight junction proteins and penetrate gut epithelial monolayers to sample bacteria. *Nat Immunol,* 2, 361-7.

SHEN, R., KAPPES, J. C., SMYTHIES, L. E., RICHTER, H. E., NOVAK, L. & SMITH, P. D. 2014. Vaginal myeloid dendritic cells transmit founder HIV-1. *J Virol,* 88, 7683-8.

SIXBEY, J. W. & YAO, Q. Y. 1992. Immunoglobulin A-induced shift of Epstein-Barr virus tissue tropism. *Science,* 255, 1578-80.

SOILLEUX, E. J., MORRIS, L. S., LESLIE, G., CHEHIMI, J., LUO, Q., LEVRONEY, E., TROWSDALE, J., MONTANER, L. J., DOMS, R. W., WEISSMAN, D., COLEMAN, N. & LEE, B. 2002. Constitutive and induced expression of DC-SIGN on dendritic cell and macrophage subpopulations *in situ* and *in vitro. J Leukoc Biol,* 71, 445-57.

SPEAR, P. G. 2004. Herpes simplex virus: receptors and ligands for cell entry. *Cell Microbiol,* 6, 401-10.

SPECK, P., HAAN, K. M. & LONGNECKER, R. 2000. Epstein-Barr virus entry into cells. *Virology,* 277, 1-5.

SPILLER, O. B. & MORGAN, B. P. 1998. Antibody-independent activation of the classical complement pathway by cytomegalovirus-infected fibroblasts. *J Infect Dis,* 178, 1597-603.

STEVCEVA, L., YOON, V., CARVILLE, A., PACHECO, B., SANTOSUOSSO, M., KORIOTH-SCHMITZ, B., MANSFIELD, K. & POZNANSKY, M. C. 2008. The efficacy of T cell-mediated immune responses is reduced by the envelope protein of the chimeric HIV-1/SIV-KB9 virus *in vivo. J Immunol,* 181, 5510-21.

SVERDLOV, E. D. 2000. Retroviruses and primate evolution. *Bioessays,* 22, 161-71.

Subject Index

A

Acquired immunodeficiency syndrome (AIDS) 14, 16, 75, 79, 99, 100, 103, 105, 108, 112
Acute EBV infection 84
Acute HIV infection 105
Acute SHIV infection 127
Adenoviral vectors 87
Adenovirus 109
Adjuvanted protein vaccines 49, 55
Adoptive transfer 37, 45, 46, 53
AIDS-associated opportunistic infections 105
Alphaherpesviridae 11, 25, 61, 122
Alphavirus-vectored subunit vaccines 46
Alphavirus vectors 50
Amplification loop 117, 119, 120, 121, 122
Antibody responses 28, 46, 48, 49, 50, 102, 106, 108
Antigen-presenting cells (APCs) 43, 55
Antiviral drugs 37, 40
AP-1 sites 80
Apoptosis 15, 17, 66, 79, 80, 82, 83, 105, 122
Asymptomatic infection 38
Attenuated alphavirus replicons 51
Attenuated Vaccine 46, 51, 50, 55, 63, 66, 67, 127, 128

B

B-cell receptor (BCR) 80, 81
BCRFI (EBV) 79
BHRF1 (EBV) 79, 80
BZLFI (EBV) 17, 77, 80, 81
BL survival 81, 82
B-lymphocytes 72, 73, 74, 78, 80
Bornavirus 3, 6
Burkitt's lymphoma (BL) 9, 16, 71, 72, 73, 74, 81, 82, 88

C

Canarypox virus 50
Capsid 10, 11, 14, 17, 26, 28, 29, 63, 64, 76, 100, 103
CD4+T cells 3, 5
CD4 receptor 101, 102, 125
CD8+ cells 32, 39
CD8+ T-cells 51, 75
CD27neg cells 44
CD34+ hematopoetic stem cells 15
CD70-expressing cells in vivo 44
Cell exhaustion 32, 25

Cell-free infectious VZV 66
Cell infection 106, 122
Cell membrane 31, 65, 101
Cell responses 5, 32, 45, 46, 49, 67, 68, 105, 109, 124
Cellular immune responses 37, 38, 47, 48, 51, 52, 53, 84
Chemokines 14, 18
Chicken pox 12, 61, 62, 63, 66, 67, 69, 85
Childhood varicella vaccination 67
Chinese hamster ovary (CHO) 49
Chinese hamster ovary cells 86
Chronic viral infection 6, 9, 117
Clinical trial evaluation 50, 51
CMV disease 38, 39, 40, 47, 48, 53, 54
CMV DNA vaccines 52
CMV infection 5, 14, 37, 38, 40, 41, 42, 43, 44, 45, 46, 53
CMV latency 42, 118, 129
CMV proteins 53, 54
CMV reactivation 39, 44, 45
CMV-seronegative 48, 49, 51
CMV-seropositive donors 45, 51, 52, 54
CMV vaccine development 46, 50
CMV vaccines 37, 46, 47, 49, 55
Complement cascade 106, 118, 119, 120, 121, 122, 123, 125
Concatamers 9
Congenital CMV infection 37, 46, 47, 54, 55
Congenital infections 46
Congenital transmission 37, 38
Convertase 118, 119, 120, 121, 123, 124
Cottontop tamarins 86, 87
Cranial root ganglia 61
CTL responses 50, 52, 55, 105, 109
C-type lectin 106, 124
Cutaneous epithelial cells 62
Cutaneous leukocyte antigen (CLA) 62
Cytomegalovirus 14, 37, 55, 105
Cytopathic effects 108
Cytotoxic T-lymphocytes (CTLs) 53, 79, 84, 85, 104, 109

D

Dendritic cells 17, 18, 51, 66, 99, 106, 107, 124, 126
Dense bodies (DBs) 53, 54
Detecting HIV infection 104
Disulfide bonds 101, 102

www.ingramcontent.com/pod-product-compliance
Lightning Source LLC
Chambersburg PA
CBHW041714210326
41598CB00007B/653